ENVIRONMENTAL ENGINEERING AND ACTIVATED SLUDGE PROCESSES

Models, Methodologies, and Applications

ENVIRONMENTAL ENGINEERING AND ACTIVATED SLUDGE PROCESSES

Models, Methodologies, and Applications

Edited by
Olga Sánchez, PhD

Apple Academic Press Inc. | Apple Academic Press Inc.
3333 Mistwell Crescent | 9 Spinnaker Way
Oakville, ON L6L 0A2 | Waretown, NJ 08758
Canada | USA

©2016 by Apple Academic Press, Inc.

First issued in paperback 2021

Exclusive worldwide distribution by CRC Press, a member of Taylor & Francis Group
No claim to original U.S. Government works

ISBN 13: 978-1-77463-708-1 (pbk)
ISBN 13: 978-1-77188-388-7 (hbk)

Library and Archives Canada Cataloguing in Publication

Environmental engineering and activated sludge processes: models, methodologies, and applications/edited by Olga Sánchez, PhD.

Includes bibliographical references and index.
Issued in print and electronic formats.
ISBN 978-1-77188-388-7 (hardcover).--ISBN 978-1-77188-389-4 (html)

1. Sewage--Purification--Activated sludge process--Case studies.
2. Sewage--Purification--Activated sludge process--Mathematical models--Case studies.
3. Sewage--Microbiology--Case studies. 4. Environmental engineering--Case studies. I. Sánchez, Olga, author, editor

TD756.E58 2016   628.3'54   C2016-900185-7   C2016-900186-5

Library of Congress Cataloging-in-Publication Data

Names: Sánchez, Olga editor.
Title: Environmental engineering and activated sludge processes : models, methodologies, and applications / [edited by] Olga Sánchez, PhD.
Description: Oakville, ON ; Waretown, NJ : Apple Academic Press, 2016. |
Includes bibliographical references and index.
Identifiers: LCCN 2016000591 (print) | LCCN 2016011493 (ebook) | ISBN 9781771883887 (hardcover : alk. paper) | ISBN 9781771883894 (eBook) | ISBN 9781771883894 ()
Subjects: LCSH: Sewage--Purification--Activated sludge process | Water--Purification--Biological treatment.
Classification: LCC TD756 .E68 2016 (print) | LCC TD756 (ebook) | DDC 628.3/54--dc23
LC record available at http://lccn.loc.gov/2016000591

About the Editor

Olga Sánchez, PhD

Olga Sánchez received her PhD degree in Biological Sciences from the Universitat Autònoma de Barcelona (Spain) in 1996. Her research began with the study of the physiology of photosynthetic bacteria and went on to the utilization of complex microbial biofilms in packed reactors for the treatment of contaminated effluents. In 2007 she became an aggregate teacher at the Department of Genetics and Microbiology of the Universitat Autònoma de Barcelona, and presently, her investigation focuses on the application of molecular techniques for the characterization of the diversity of different natural microbial communities, such as marine environments or wastewater treatment systems. These methodologies include clone libraries, FISH (Fluorescence *In Situ* hybridization), fingerprinting techniques such as DGGE (Denaturing Gradient Gel Electrophoresis), and next-generation sequencing technologies. An active researcher, she has authored and coauthored more than forty research publications in the field of microbial ecology. She has also edited other compendium volumes.

Contents

Acknowledgment and How to Cite .... *xi*

List of Contributors .... *xiii*

Introduction .... *xix*

Part I: Models and Kinetics

1. **Experimental and Theoretical Approaches for the Surface Interaction between Copper and Activated Sludge Microorganisms at Molecular Scale** 3

   Hong-Wei Luo, Jie-Jie Chen, Guo-Ping Sheng, Ji-Hu Su, Shi-Qiang Wei, and Han-Qing Yu

2. **A Novel Protocol for Model Calibration in Biological Wastewater Treatment** 23

   Ao Zhu, Jianhua Guo, Bing-Jie Ni, Shuying Wang, Qing Yang, and Yongzhen Peng

3. **Sorption and Release of Organics by Primary, Anaerobic, and Aerobic Activated Sludge Mixed with Raw Municipal Wastewater** 49

   Oskar Modin, Soroush Saheb Alam, Frank Persson, and Britt-Marie Wilén

4. **Removal Mechanisms and Kinetics of Trace Tetracycline by Two Types of Activated Sludge Treating Freshwater Sewage and Saline Sewage** 75

   Bing Li and Tong Zhang

Part II: Diversity of Activated Sludge and Process Microbiology

5. **Evaluation of Simultaneous Nutrient and COD Removal with Polyhydroxybutyrate (PHB) Accumulation Using Mixed Microbial Consortia under Anoxic Condition and Their Bioinformatics Analysis** 105

   Jyotsnarani Jena, Ravindra Kumar, Anshuman Dixit, Sony Pandey, and Trupti Das

6. **Characterization of Pure Cultures Isolated from
    Sulfamethoxazole-acclimated Activated Sludge with Respect to
    Taxonomic Identification and Sulfamethoxazole Biodegradation
    Potential** 137
    Bastian Herzog, Hilde Lemmer, Harald Horn, and Elisabeth Müller

7. **Assessing Bacterial Diversity in a Seawater-processing
    Wastewater Treatment Plant by 454-pyrosequencing of the
    16S rRNA and *amoA* Genes** 159

    Olga Sánchez, Isabel Ferrera, Jose M. González, and Jordi Mas

Part III: Nitrogen and Phosphorus Removal

8. **An Efficient Process for Wastewater Treatment to Mitigate
    Free Nitrous Acid Generation and its Inhibition on Biological
    Phosphorus Removal** 175

    Jianwei Zhao, Dongbo Wang, Xiaoming Li, Qi Yang, Hongbo Chen,
    Yu Zhong, Hongxue An, and Guangming Zeng

Part IV: Xenobiotics

9. **Bacterial Consortium and Axenic Cultures Isolated from
    Activated Sewage Sludge for Biodegradation of Imidazolium-
    based Ionic Liquid** 201

    Marta Markiewicz, Joanna Henke, Anna Brillowska-Dabrowska,
    Stefan Stolte, Justyna Luczak, and Christian Jungnickel

Part V: Methane Production

10. **Zero Valent Iron Significantly Enhances Methane Production
    from Waste Activated Sludge by Improving Biochemical
    Methane Potential Rather Than Hydrolysis Rate** 219
    Yiwen Liu, Qilin Wang, Yaobin Zhang, and Bing-Jie Ni

11. **Towards a Metagenomic Understanding on Enhanced
    Biomethane Production from Waste Activated Sludge after
    pH 10 Pretreatment** 237

    Mabel Ting Wong, Dong Zhang, Jun Li, Raymond Kin Hi Hui,
    Hein Min Tun, Manreetpal Singh Brar, Tae-Jin Park, Yinguang Chen, and
    Frederick C. Leung

Part VI: Biofilm Bioreactors

12. **Isolation and Molecular Characterization of Biofouling Bacteria and Profiling of Quorum Sensing Signal Molecules from Membrane Bioreactor Activated Sludge** 269

    Harshad Lade, Diby Paul, and Ji Hyang Kweon

13. **Scenario Analysis of Nutrient Removal from Municipal Wastewater by Microalgal Biofilms** 293

    Nadine C. Boelee, Hardy Temmink, Marcel Janssen, Cees J. N. Buisman, and René H. Wijffels

Author Notes 311

Index 319

Acknowledgment and How to Cite

The editor and publisher thank each of the authors who contributed to this book. The chapters in this book were previously published elsewhere. To cite the work contained in this book and to view the individual permissions, please refer to the citation at the beginning of each chapter. Each chapter was carefully selected by the editor; the result is a book that looks at activated sludge processes from a variety of perspectives. The chapters included are broken into six sections, which describe the following topics:

- The main objective of Chapter 1 is to get a deep insight into the microscopic-level interaction between heavy metal Cu(II) and activated sludge, as well as its extracellular polymeric substances.
- Chapter 2 develops a novel systematical approach—the Numerical Optimal Approaching Procedure (NOAP)—for efficient calibration and validation of activated sludge models.
- The goal of Chapter 3 is to quantify the sorption of particulate and dissolved organic compounds onto different types of sludge available at wastewater treatment plants.
- Chapter 4 systematically examines the elimination of tetracycline by two types of activated sludge, treating freshwater sewage and saline sewage respectively, at environmentally relevant concentrations.
- In Chapter 5, a set of batch experiments was designed to study the potential of enriched microbial consortia for simultaneous nutrient removal and PHB accumulation under complete anoxic conditions with varying initial carbon load.
- Chapter 6's emphasis is on clarifying the effect in which the addition of readily degradable carbon and/or nitrogen sources in some cases significantly enhanced sulfamethoxazole elimination, while in other cases supplementation showed no effect.
- Chapter 7 analyzes the 16S rRNA and *amoA* genes diversity in an activated sludge of a wastewater treatment plant, with the particularity to utilize seawater.
- The purpose of Chapter 8 is to report an efficient method for significantly mitigating the generation of free nitrous acid and its inhibition on polyphosphate accumulating organisms.

- In the course of Chapter 9, nine strains of bacteria from adapted activated sewage sludge were isolated, and the influence of 1-methyl-3-octylimidazolium chloride on their growth was tested.
- In Chapter 10, the impacts of three different types of zero valent iron on methane production from water activated sludge in anaerobic digestion were evaluated systematically using both experimental and mathematical approaches.
- In Chapter 11, a shotgun metagenomic approach was chosen to study potential shifts in microbial communities and/or gene contents that could help explain elevated productions of methane under the authors' novel pretreatment method.
- The objectives of Chapter 12 were to characterize the key bacterial community responsible for membrane biofouling in a membrane bioreactor treating wastewater, and profiling of the quorum sensing signal molecules involved.
- Chapter 13 aims to get insight in the feasibility of using microalgal biofilms for municipal wastewater treatment.

List of Contributors

Hongxue An
College of Environmental Science and Engineering, Hunan University, Changsha 410082, China; Key Laboratory of Environmental Biology and Pollution Control, Hunan University, Ministry of Education, Changsha 410082, China

Nadine C. Boelee
Wetsus—Centre of Excellence for Sustainable Water Technology, P.O. Box 1113, Leeuwarden 8900 CC, The Netherlands; Sub-department of Environmental Technology, Wageningen University, P.O. Box 8129, Wageningen 6700 EV, The Netherlands; Bioprocess Engineering, Wageningen University, P.O. Box 8129, Wageningen 6700 EV,The Netherlands

Anna Brillowska-Dabrowska
Department of Microbiology, Chemical Faculty, Gdánsk University of Technology, ul. Narutowicza 11/12, 80-233 Gdánsk, Poland

Cees J. N. Buisman
Wetsus—Centre of Excellence for Sustainable Water Technology, P.O. Box 1113, Leeuwarden 8900 CC, The Netherlands; Sub-department of Environmental Technology, Wageningen University, P.O. Box 8129, Wageningen 6700 EV, The Netherlands

Hongbo Chen
College of Environmental Science and Engineering, Hunan University, Changsha 410082, China; Key Laboratory of Environmental Biology and Pollution Control, Hunan University, Ministry of Education, Changsha 410082, China

Jie-Jie Chen
Department of Modern Physics, University of Science and Technology of China, Hefei, 230026, China

Yinguang Chen
State Key Laboratory of Pollution Control and Resources Reuse, School of Environmental Science and Engineering, Tongji University, 1239 Siping Road, Shanghai 200092, China

Trupti Das
CSIR-Institute of Minerals and Materials Technology, Bhubaneswar, Odisha, India

Anshuman Dixit
Institute of Life Sciences, Bhubaneswar, Bhubaneswar, Odisha, India

Isabel Ferrera
Departament de Biologia Marina i Oceanografia, Institut de Ciències del Mar, ICM-CSIC, Barcelona, Spain

Jose M. González
Department of Microbiology, University of La Laguna, ES-38206 La Laguna, Tenerife, Spain

Jianhua Guo
Key Laboratory of Beijing for Water Quality Science and Water Environmental Recovery Engineering, Engineering Research Center of Beijing, Beijing University of Technology, Beijing 100124, PR China; Advanced Water Management Centre (AWMC), The University of Queensland, St Lucia, Brisbane, QLD 4072, Australia

Joanna Henke
Department of Chemical Technology, Chemical Faculty, Gdánsk University of Technology, ul. Narutowicza 11/12, 80-233 Gdánsk, Poland

Bastian Herzog
Urban Water Systems Engineering, Technische Universität München, Am Coulombwall, D-85748 Garching, Germany

Harald Horn
Karlsruhe Institute of Technology, Engler-Bunte-Institut, Bereich Wasserchemie und Wassertechnologie, D-76131 Karlsruhe, Germany

Marcel Janssen
Bioprocess Engineering, Wageningen University, P.O. Box 8129, Wageningen 6700 EV, The Netherlands

Jyotsnarani Jena
CSIR-Institute of Minerals and Materials Technology, Bhubaneswar, Odisha, India

Christian Jungnickel
Department of Chemical Technology, Chemical Faculty, Gdánsk University of Technology, ul. Narutowicza 11/12, 80-233 Gdánsk, Poland

Raymond Kin Hi Hui
5 N01, Kadoorie Biological Sciences Building, School of Biological Sciences, The University of Hong Kong, Pokfulam Road, Hong Kong,

Ravindra Kumar
Institute of Life Sciences, Bhubaneswar, Bhubaneswar, Odisha, India

Ji Hyang Kweon
Department of Environmental Engineering, Konkuk University, Seoul-143-701, Korea

Harshad Lade
Department of Environmental Engineering, Konkuk University, Seoul-143-701, Korea

Hilde Lemmer
Bavarian Environment Agency, Bürgermeister-Ulrich-Str. 160, D-86179 Augsburg, Germany

Frederick C. Leung
5 N01, Kadoorie Biological Sciences Building, School of Biological Sciences, The University of Hong Kong, Pokfulam Road, Hong Kong, Hong Kong; Bioinformatics Center, Nanjing Agricultural University, Nanjing, China

Bing Li
Environmental Biotechnology Laboratory, Department of Civil Engineering, The University of Hong Kong, Pokfulam Road, Hong Kong, SAR, China

Jun Li
5 N01, Kadoorie Biological Sciences Building, School of Biological Sciences, The University of Hong Kong, Pokfulam Road, Hong Kong,

Xiaoming Li
College of Environmental Science and Engineering, Hunan University, Changsha 410082, China; Key Laboratory of Environmental Biology and Pollution Control, Hunan University, Ministry of Education, Changsha 410082, China

Yiwen Liu
Advanced Water Management Centre, The University of Queensland, St Lucia, QLD 4072, Australia

Hong-Wei Luo
CAS Key Laboratory of Urban Pollutant Conversion, Department of Chemistry

Justyna Łuczak
Department of Chemical Technology, Chemical Faculty, Gdánsk University of Technology, ul. Narutowicza 11/12, 80-233 Gdánsk, Poland

Marta Markiewicz
Department of Chemical Technology, Chemical Faculty, Gdánsk University of Technology, ul. Narutowicza 11/12, 80 233 Gdánsk, Poland

Jordi Mas
Departament de Genètica i Microbiologia, Universitat Autònoma de Barcelona, 08193 Bellaterra, Spain

Oskar Modin
Division of Water Environment Technology, Department of Civil and Environmental Engineering, Chalmers University of Technology, Gothenburg, Sweden

Elisabeth Müller
Urban Water Systems Engineering, Technische Universität München, Am Coulombwall, D-85748 Garching, Germany

Bing-Jie Ni
Advanced Water Management Centre (AWMC), The University of Queensland, St Lucia, Brisbane, QLD 4072, Australia

Sony Pandey
CSIR-Institute of Minerals and Materials Technology, Bhubaneswar, Odisha, India

Tae-Jin Park
5 N01, Kadoorie Biological Sciences Building, School of Biological Sciences, The University of Hong Kong, Pokfulam Road, Hong Kong, Hong Kong

Diby Paul
Department of Environmental Engineering, Konkuk University, Seoul-143-701, Korea

Yongzhen Peng
Key Laboratory of Beijing for Water Quality Science and Water Environmental Recovery Engineering, Engineering Research Center of Beijing, Beijing University of Technology, Beijing 100124, PR China

Frank Persson
Division of Water Environment Technology, Department of Civil and Environmental Engineering, Chalmers University of Technology, Gothenburg, Sweden

Soroush Saheb Alam
Division of Water Environment Technology, Department of Civil and Environmental Engineering, Chalmers University of Technology, Gothenburg, Sweden

Olga Sánchez
Departament de Genètica i Microbiologia, Universitat Autònoma de Barcelona, 08193 Bellaterra, Spain

Guo-Ping Sheng
CAS Key Laboratory of Urban Pollutant Conversion, Department of Chemistry

Manreetpal Singh Brar
5 N01, Kadoorie Biological Sciences Building, School of Biological Sciences, The University of Hong Kong, Pokfulam Road, Hong Kong

Stefan Stolte
Center for Environmental Research and Sustainable Technology, University of Bremen, UFT Leobener Strasse, 28359 Bremen, Germany

Ji-Hu Su
Department of Modern Physics, University of Science and Technology of China, Hefei, 230026, China

Hardy Temmink
Wetsus—Centre of Excellence for Sustainable Water Technology, P.O. Box 1113, Leeuwarden 8900 CC, The Netherlands; Sub-department of Environmental Technology, Wageningen University, P.O. Box 8129, Wageningen 6700 EV, The Netherlands

Hein Min Tun
5 N01, Kadoorie Biological Sciences Building, School of Biological Sciences, The University of Hong Kong, Pokfulam Road, Hong Kong

Dongbo Wang
College of Environmental Science and Engineering, Hunan University, Changsha 410082, China; Key Laboratory of Environmental Biology and Pollution Control, Hunan University, Ministry of Education, Changsha 410082, China; State Key Laboratory of Pollution Control and Resources Reuse, School of Environmental Science and Engineering, Tongji University, 1239 Siping Road, Shanghai 200092, China; Advanced Water Management Centre, The University of Queensland, QLD 4072, Australia; Jiangsu Tongyan Environmental Production Science and Technology Co. Ltd., Yancheng, 224000, China

Qilin Wang
Advanced Water Management Centre, The University of Queensland, St Lucia, QLD 4072, Australia

Shuying Wang
Key Laboratory of Beijing for Water Quality Science and Water Environmental Recovery Engineering, Engineering Research Center of Beijing, Beijing University of Technology, Beijing 100124, PR China

Shi-Qiang Wei
National Synchrotron Radiation Laboratory, University of Science and Technology of China, Hefei, 230029, China

René H. Wijffels
Bioprocess Engineering, Wageningen University, P.O. Box 8129, Wageningen 6700 EV,The Netherlands

Britt-Marie Wilén
Division of Water Environment Technology, Department of Civil and Environmental Engineering, Chalmers University of Technology, Gothenburg, Sweden

Mabel Ting Wong
5 N01, Kadoorie Biological Sciences Building, School of Biological Sciences, The University of Hong Kong, Pokfulam Road, Hong Kong, Hong Kong

Qi Yang
College of Environmental Science and Engineering, Hunan University, Changsha 410082, China; Key Laboratory of Environmental Biology and Pollution Control, Hunan University, Ministry of Education, Changsha 410082, China

Qing Yang
Key Laboratory of Beijing for Water Quality Science and Water Environmental Recovery Engineering, Engineering Research Center of Beijing, Beijing University of Technology, Beijing 100124, PR China

Han-Qing Yu
CAS Key Laboratory of Urban Pollutant Conversion, Department of Chemistry

Guangming Zeng
College of Environmental Science and Engineering, Hunan University, Changsha 410082, China; Key Laboratory of Environmental Biology and Pollution Control, Hunan University, Ministry of Education, Changsha 410082, China

Dong Zhang
State Key Laboratory of Pollution Control and Resources Reuse, School of Environmental Science and Engineering, Tongji University, 1239 Siping Road, Shanghai 200092, China

Tong Zhang
Environmental Biotechnology Laboratory, Department of Civil Engineering, The University of Hong Kong, Pokfulam Road, Hong Kong, SAR, China

Yaobin Zhang
Key Laboratory of Industrial Ecology and Environmental Engineering, Ministry of Education, School of Environmental Science and Technology, Dalian University of Technology, Dalian 116024, China

Jianwei Zhao
College of Environmental Science and Engineering, Hunan University, Changsha 410082, China; Key Laboratory of Environmental Biology and Pollution Control, Hunan University, Ministry of Education, Changsha 410082, China

Yu Zhong
College of Environmental Science and Engineering, Hunan University, Changsha 410082, China; Key Laboratory of Environmental Biology and Pollution Control, Hunan University, Ministry of Education, Changsha 410082, China

Ao Zhu
Key Laboratory of Beijing for Water Quality Science and Water Environmental Recovery Engineering, Engineering Research Center of Beijing, Beijing University of Technology, Beijing 100124, PR China; Tsinghua Holding Human Settlements Environment Institute, Beijing 100083, PR China

Introduction

Activated sludge constitutes one of the most important biotechnological processes for wastewater treatment and environmental protection. In this suspended-growth process, a complex mixture of microorganisms has the ability to degrade organic matter, remove nutrients and transform toxic compounds into harmless products. For this reason, it is crucial to understand the microbial community structure and the processes that underlie behind activated sludge.

Different molecular approaches have been applied during the last decades for characterizing the structure and function of these microbial communities, and the key microorganisms for the major treatment processes such as nitrification, denitrification, or biological phosphorus removal have been identified. However, to obtain a deeper understanding of such systems, it is necessary to integrate these tools with ecological theory, and to develop conceptual and predictive mathematical models to describe the main activities and how the populations can function, including their interactions, their roles and their implications for the operation of such ecosystems. These models can constitute a useful framework for proposing and testing theories for the future design and management of wastewater treatment systems.

The articles in this compendium were chosen to give an overview of the most important aspects that highlight the importance of activated sludge for microbiologists and engineers, emphasizing its role as an effective tool for contaminants removal.

Dr. Olga Sánchez

Interactions between metals and activated sludge microorganisms substantially affect the speciation, immobilization, transport, and bioavailability of trace heavy metals in biological wastewater treatment plants. In Chapter 1, the interaction of Cu(II), a typical heavy metal, onto activated sludge microorganisms was studied in-depth using a multi-technique approach. The complexing structure of Cu(II) on microbial surface was revealed by X-ray absorption fine structure (XAFS) and electron paramagnetic resonance (EPR) analysis. EPR spectra indicated that Cu(II) was held in inner-sphere surface complexes of octahedral coordination with tetragonal distortion of axial elongation. XAFS analysis further suggested that the surface complexation between Cu(II) and microbial cells was the distorted inner-sphere coordinated octahedra containing four short equatorial bonds and two elongated axial bonds. To further validate the results obtained from the XAFS and EPR analysis, density functional theory calculations were carried out to explore the structural geometry of the Cu complexes. These results are useful to better understand the speciation, immobilization, transport, and bioavailability of metals in biological wastewater treatment plants.

Activated sludge models (ASMs) have been widely used for process design, operation and optimization in wastewater treatment plants. However, it is still a challenge to achieve an efficient calibration for reliable application by using the conventional approaches. Chapter 2 proposes a novel calibration protocol, i.e. Numerical Optimal Approaching Procedure (NOAP), for the systematic calibration of ASMs. The NOAP consists of three key steps in an iterative scheme flow: i) global factors sensitivity analysis for factors fixing; ii) pseudo-global parameter correlation analysis for non-identifiable factors detection; and iii) formation of a parameter subset through an estimation by using genetic algorithm. The validity and applicability are confirmed using experimental data obtained from two independent wastewater treatment systems, including a sequencing batch reactor and a continuous stirred-tank reactor. The results indicate that the NOAP can effectively determine the optimal parameter subset and successfully perform model calibration and validation for these two different systems. The proposed NOAP is expected to use for automatic calibration of ASMs and be applied potentially to other ordinary differential equations models.

New activated sludge processes that utilize sorption as a major mechanism for organics removal are being developed to maximize energy recovery from wastewater organics, or as enhanced primary treatment technologies. To model and optimize sorption-based activated sludge processes, further knowledge about sorption of organics onto sludge is needed. Chapter 3 compared primary-, anaerobic-, and aerobic activated sludge as sorbents, determined sorption capacity and kinetics, and investigated some characteristics of the organics being sorbed. Batch sorption assays were carried out without aeration at a mixing velocity of 200 rpm. Only aerobic activated sludge showed net sorption of organics. Sorption of dissolved organics occurred by a near-instantaneous sorption event followed by a slower process that obeyed 1st order kinetics. Sorption of particulates also followed 1st order kinetics but there was no instantaneous sorption event; instead there was a release of particles upon mixing. The 5-min sorption capacity of activated sludge was 6.5 ± 10.8 mg total organic carbon (TOC) per g volatile suspend solids (VSS) for particulate organics and 5.0 ± 4.7 mgTOC/gVSS for dissolved organics. The observed instantaneous sorption appeared to be mainly due to organics larger than 20 kDa in size being sorbed, although molecules with a size of about 200 Da with strong UV absorbance at 215–230 nm were also rapidly removed.

Understanding the removal mechanisms and kinetics of trace tetracycline by activated sludge is critical to both evaluation of tetracycline elimination in sewage treatment plants and risk assessment/management of tetracycline released to soil environment due to the application of biosolids as fertilizer. In Chapter 4, adsorption is found to be the primary removal mechanism while biodegradation, volatilization, and hydrolysis can be ignored in this study. Adsorption kinetics was well described by pseudo-second-order model. Faster adsorption rate ($k_2=2.04\times10^{-2}$gmin$^{-1}\mu$g^{-1}) and greater adsorption capacity ($q_e=38.8$ μgg^{-1}) were found in activated sludge treating freshwater sewage. Different adsorption rate and adsorption capacity resulted from chemical properties of sewage matrix rather than activated sludge surface characteristics. The decrease of tetracycline adsorption in saline sewage was mainly due to Mg^{2+} which significantly reduced adsorption distribution coefficient (K_d) from $12,990\pm260$ to $4,690\pm180$ Lkg^{-1}. Species-specific adsorption distribution coefficients followed the order of $K_d^{=00} >> K_d^{=-0} > K_d^{=-0}$. Contribution of zwitterionic tetracycline to the overall adsorption was >90 % in the actual pH range in aeration tank. Adsorption of tetracycline in a wide range of temperature (10 to 35 °C) followed the Freundlich adsorption isotherm well.

In Chapter 5, simultaneous nitrate-N, phosphate and COD removal was evaluated from synthetic waste water using mixed microbial consortia in an anoxic environment under various initial carbon load (ICL) in a batch scale reactor system. Within 6 hours of incubation, enriched DN-PAOs (Denitrifying Polyphosphate Accumulating Microorganisms) were able to remove maximum COD (87%) at 2g/L of ICL whereas maximum nitrate-N (97%) and phosphate (87%) removal along with PHB accumulation (49 mg/L) was achieved at 8 g/L of ICL. Exhaustion of nitrate-N, beyond 6 hours of incubation, had a detrimental effect on COD and phosphate removal rate. Fresh supply of nitrate-N to the reaction medium, beyond 6 hours, helped revive the removal rates of both COD and phosphate. Therefore, it was apparent that in spite of a high carbon load, maximum COD and nutrient removal can be maintained, with adequate nitrate-N availability. Denitrifying condition in the medium was evident from an increasing pH trend. PHB accumulation by the mixed culture was directly proportional to ICL; however the time taken for accumulation at higher ICL was more. Unlike conventional EBPR, PHB depletion did not support phosphate accumulation in this case. The unique aspect of all the batch studies were PHB accumulation was observed along with phosphate uptake and nitrate reduction under anoxic conditions. Bioinformatics analysis followed by pyrosequencing of the mixed culture DNA from the seed sludge revealed the dominance of denitrifying population, such as *Corynebacterium, Rhodocyclus* and *Paraccocus* (*Alphaproteobacteria* and *Betaproteobacteria*). Rarefaction curve indicated complete bacterial population and corresponding number of OTUs through sequence analysis. Chao1 and Shannon index (H') was used to study the diversity of sampling. "UCI95" and "LCI95" indicated 95% confidence level of upper and lower values of Chao1 for each distance. Values of Chao1 index supported the results of rarefaction curve.

Sulfamethoxazole (SMX, sulfonamide antibiotic) biodegradation by activated sludge communities (ASC) is still only partly understood. Chapter 6 focuses on nine different bacteria species capable of SMX biodegradation that were isolated from SMX-acclimated ASC. Initially 110 pure cultures, isolated from activated sludge, were screened by UV-absorbance measurements (UV-AM) for their SMX biodegradation potential. Identification via almost complete 16S rRNA gene sequencing revealed five

Pseudomonas spp., one *Brevundimonas* sp., one *Variovorax* sp. and two *Microbacterium* spp.. Thus seven species belonged to the phylum *Proteobacteria* and two to *Actinobacteria*. These cultures were subsequently incubated in media containing 10 mg L^{-1} SMX and different concentrations of carbon (sodium-acetate) and nitrogen (ammonium-nitrate). Different biodegradation patterns were revealed with respect to media composition and bacterial species. Biodegradation, validated by LC-UV measurements to verify UV-AM, occurred very fast with 2.5 mg L^{-1} d^{-1} SMX being biodegraded in all pure cultures in, for UV-AM modified, R2A-UV medium under aerobic conditions and room temperature. However, reduced and different biodegradation rates were observed for setups with SMX provided as co-substrate together with a carbon/nitrogen source at a ratio of DOC:N – 33:1 with rates ranging from 1.25 to 2.5 mg L^{-1} d^{-1}. Media containing only SMX as carbon and nitrogen source proved the organisms' ability to use SMX as sole nutrient source where biodegradation rates decreased to 1.0 – 1.7 mg L^{-1} d^{-1}. The different taxonomically identified species showed specific biodegradation rates and behaviours at various nutrient conditions. Readily degradable energy sources seem to be crucial for efficient SMX biodegradation.

The bacterial community composition of activated sludge from a wastewater treatment plant (Almería, Spain) with the particularity of using seawater was investigated by applying 454-pyrosequencing in Chapter 7. The results showed that *Deinococcus-Thermus*, *Proteobacteria*, *Chloroflexi* and *Bacteroidetes* were the most abundant retrieved sequences, while other groups, such as *Actinobacteria*, *Chlorobi*, *Deferribacteres*, *Firmicutes*, *Planctomycetes*, *Spirochaetes* and *Verrumicrobia* were reported at lower proportions. Rarefaction analysis showed that very likely the diversity is higher than what could be described despite most of the unknown microorganisms probably correspond to rare diversity. Furthermore, the majority of taxa could not be classified at the genus level and likely represent novel members of these groups. Additionally, the nitrifiers in the sludge were characterized by pyrosequencing the *amoA* gene. In contrast, the nitrifying bacterial community, dominated by the genera *Nitrosomonas*, showed a low diversity and rarefaction curves exhibited saturation. These results suggest that only a few populations of low abundant but specialized bacteria are responsible for removal of ammonia in these saline wastewater systems.

Free nitrous acid (FNA), which is the protonated form of nitrite and inevitably produced during biological nitrogen removal, has been demonstrated to strongly inhibit the activity of polyphosphate accumulating organisms (PAOs). Chapter 8 reports an efficient process for wastewater treatment, i.e., the oxic/anoxic/ oxic/extended-idle process to mitigate the generation of FNA and its inhibition on PAOs. The results showed that this new process enriched more PAOs which thereby achieved higher phosphorus removal efficiency than the conventional four-step (i.e., anaerobic/oxic/anoxic/oxic) biological nutrient removal process ($41\pm7\%$ versus $30\pm5\%$ in abundance of PAOs and $97\pm0.73\%$ versus $82\pm1.2\%$ in efficiency of phosphorus removal). It was found that this new process increased pH value but decreased nitrite accumulation, resulting in the decreased FNA generation. Further experiments showed that the new process could alleviate the inhibition of FNA on the metabolisms of PAOs even under the same FNA concentration.

Extensive research and increasing number of potential industrial applications made ionic liquids (ILs) important materials in design of new, cleaner technologies. Together with the technological applicability, the environmental fate of these chemicals is considered and significant efforts are being made in designing strategies to mitigate their potential negative impacts. Many ILs are proven to be poorly biodegradable and relatively toxic. Bioaugmentation is known as one of the ways of enhancing the microbial capacity to degrade xenobiotics by addition of specialized strains. The aim of Chapter 9 was to select microbial species that could be used for bioaugmentation in order to enhance biodegradation of ILs in the environment. The authors subjected activated sewage sludge to the selective pressure of 1-methyl-3-octylimidazolium chloride ([OMIM][Cl]) and isolated nine strains of bacteria which were able to prevail in these conditions. Subsequently, the authors utilized axenic cultures (pure cultures) of these bacteria as well as mixed consortium to degrade this IL. In addition, they performed growth inhibition tests and found that bacteria were able to grow in 2 mM, but not in 20 mM solutions of [OMIM][Cl]. The biodegradation conducted by the isolated consortium was higher than conducted by the activated sewage sludge when normalized by the cell density, which indicates that the isolated strains seem specifically suited to degrade the IL.

Anaerobic digestion has been widely applied for waste activated sludge (WAS) treatment. However, methane production from anaerobic digestion of WAS is usually limited by the slow hydrolysis rate and/or poor biochemical methane potential of WAS. Chapter 10 systematically studied the effects of three different types of zero valent iron (i.e., iron powder, clean scrap and rusty scrap) on methane production from WAS in anaerobic digestion, by using both experimental and mathematical approaches. The results demonstrated that both the clean and the rusty iron scrap were more effective than the iron powder for improving methane production from WAS. Model-based analysis showed that ZVI addition significantly enhanced methane production from WAS through improving the biochemical methane potential of WAS rather than its hydrolysis rate. Economic analysis indicated that the ZVI-based technology for enhancing methane production from WAS is economically attractive, particularly considering that iron scrap can be freely acquired from industrial waste. Based on these results, the ZVI-based anaerobic digestion process of this work could be easily integrated with the conventional chemical phosphorus removal process in wastewater treatment plant to form a cost-effective and environment-friendly approach, enabling maximum resource recovery/reuse while achieving enhanced methane production in wastewater treatment system.

Understanding the effects of pretreatment on anaerobic digestion of sludge waste from wastewater treatment plants is becoming increasingly important, as impetus moves towards the utilization of sludge for renewable energy production. Although the field of sludge pretreatment has progressed significantly over the past decade, critical questions concerning the underlying microbial interactions remain unanswered. In Chapter 11, a metagenomic approach was adopted to investigate the microbial composition and gene content contributing to enhanced biogas production from sludge subjected to a novel pretreatment method (maintaining pH at 10 for 8 days) compared to other documented methods (ultrasonic, thermal and thermal-alkaline). The results showed that pretreated sludge attained a maximum methane yield approximately 4-fold higher than that of the blank un-pretreated sludge set-up at day 17. Both the microbial and metabolic consortium shifted extensively towards enhanced biodegradation subsequent to pretreatment, providing insight for the enhanced methane yield. The prevalence of *Methanosaeta thermophila* and *Methanothermobacter thermautotrophicus*, together with the functional affiliation of enzymes-encoding genes suggested an acetoclastic and hydrogenotrophic methanogenesis pathway. Additionally, an alternative enzymology in *Methanosaeta* was observed. This study is the first to provide a microbiological understanding of improved biogas production subsequent to a novel waste sludge pretreatment method. The knowledge garnered will assist the design of more efficient pretreatment methods for biogas production in the future.

The formation of biofilm in a membrane bioreactor depends on the production of various signaling molecules like N-acyl homoserine lactones (AHLs). In Chapter 12, a total of 200 bacterial strains were isolated from membrane bioreactor activated sludge and screened for AHLs production using two biosensor systems, *Chromobacterium violaceum* CV026 and *Agrobacterium tumefaciens* A136. A correlation between AHLs production and biofilm formation has been made among screened AHLs producing strains. The 16S rRNA gene sequence analysis revealed the dominance of *Aeromonas* and *Enterobacter* sp. in AHLs production; however few a species of *Serratia, Leclercia, Pseudomonas, Klebsiella, Raoultella* and *Citrobacter* were also identified. The chromatographic characterization of sludge extract showed the presence of a broad range of quorum sensing signal molecules. Further identification of sludge AHLs by thin layer chromatography bioassay and high performance liquid chromatography confirms the presence of C4-HSL, C6-HSL, C8-HSL, 3-oxo-C8-HSL, C10-HSL, C12-HSL, 3-oxo-C12-HSL and C14-HSL. The occurrence of AHLs in sludge extract and dominance of *Aeromonas* and *Enterobacter* sp. in activated sludge suggests the key role of these bacterial strains in AHLs production and thereby membrane fouling.

Microalgae can be used for the treatment of municipal wastewater. The application of microalgal biofilms in wastewater treatment systems seems attractive, being able to remove nitrogen, phosphorus and COD from wastewater at a short hydraulic retention time. Chapter 13 therefore investigates the area requirement, achieved effluent concentrations and biomass production of a hypothetical large-scale microalgal biofilm system treating municipal wastewater. Three scenarios were defined: using microalgal biofilms: (1) as a post-treatment; (2) as a second stage of wastewater treatment, after a first stage in which COD is removed by activated sludge; and (3) in a symbiotic microalgal/heterotrophic system. The analysis showed that in the Netherlands, the area requirements for these three scenarios range from 0.32 to 2.1 m² per person equivalent. Moreover, it was found that it was not possible to simultaneously remove all nitrogen and phosphorus from the wastewater, because of the nitrogen:phosphorus ratio in the wastewater. Phosphorus was limiting in the post-treatment scenario, while nitrogen was limiting in the two other scenarios. Furthermore, a substantial amount of microalgal biomass was produced, ranging from 13 to 59 g per person equivalent per day. These findings show that microalgal biofilm systems hold large potential as seasonal wastewater treatment systems and that it is worthwhile to investigate these systems further.

PART I

MODELS AND KINETICS

CHAPTER 1

Experimental and Theoretical Approaches for the Surface Interaction between Copper and Activated Sludge Microorganisms at Molecular Scale

HONG-WEI LUO, JIE-JIE CHEN, GUO-PING SHENG, JI-HU SU, SHI-QIANG WEI, AND HAN-QING YU

1.1 INTRODUCTION

Interactions between metals and microorganisms play an important role in the geochemical cycling of trace heavy metals. In biological wastewater treatment processes, microorganisms in activated sludge have a high metal complexation capacity and then substantially affect the speciation, immobilization, transport, and bioavailability of metals in biological wastewater treatment plants. Previous study showed that there are many heavy metals (e.g., copper) in municipal and industrial wastewaters [1]. It is of great significance to study the surface complexation of metals on microorganisms

Experimental and Theoretical Approaches for the Surface Integration between Copper and Activated Sludge Microorganisms at Molecular Scale. © *Luo H-W, Chen J-J, Sheng G-P, Su J-H, Wei S-Q, and Yu H-Q.* Scientific Reports 4, 7078 (2014), doi: 10.1038/srep07078. *Used with the permission of the authors.*

and to understand their fates in biological wastewater treatment plants. In addition, extracellular polymeric substances (EPS), high-molecular weight compounds secreted by microorganisms, also have a significant impact on the fates of heavy metals [2,3]. The relevant functional groups involved in the interaction between metals and microorganisms are reported to be –COOH, –OH, –NH$_2$, and –PO$_4$, etc [4].

However, there are contradictory reports about the interaction between heavy metals and microbial cells in the presence and absence of EPS. In some studies it was found that the presence of EPS did not significantly affect the interaction [5]; while others reported that the presence of EPS had a substantial effect on the interaction [6]. Such a contradiction may be attributed to the different experimental techniques employed in these works. Furthermore, the complicated structures of activated sludge and high water content make it difficult to explore the interaction mechanisms between heavy metals and microbial cells at a microscale. Previous studies showed that activated sludge microorganisms had high complexing capability to Cu [7–9], which was particle-diffusion-controlled, and followed the pseudo-second-order rate kinetics [7]. The complexing process was also obeyed Freundlich and Langmuir models, and the saturation amount of Cu(II) adsorbed by biomass was found to be 2.00 mmol/g [9]. Although the macroscopic adsorption of many kinds of heavy metals (including Cu) onto activated sludge has been investigated intensively, and many analytical methodologies, such as potentiometric titrations, electrophoretic measurements, and infrared spectroscopy [10], have been also developed to characterize this interaction. However, the surface complexing microstructure between the heavy metals and the functional groups in the sludge surface are still not well known. So there is still a need for further investigations of the surface interaction at molecular scale.

The main objective of this study is to get a deep insight into the microscopic-level interaction between heavy metal Cu(II) and activated sludge as well as its EPS, providing the microstructure of Cu(II) complexing with the functional groups of activated sludge. Spectroscopic measurements can be used to probe surface speciation and sorption mechanisms at molecular scale. For example, electron paramagnetic resonance (EPR) is able to give the geometrical information about paramagnetic metallic cations (e.g., Cu), and qualitative and quantitative information about un-paired electrons of heavy metals after complexation. X-ray absorption fine structure (XAFS) analysis is another important means to describe the local structure including bond distance, coordination number, and type of near-neighbors surrounding a specific element [11,12], which is very sensitive to molecular complexation. The density functional theory (DFT) calcula-tion provides useful information about the interpretation of experimental spectroscopic data to identify possible coordination environments of ad-sorbed Cu [13–15]. The experimental and calculation results could pro-vide more detailed microscopic chemical structure information about the interactions between heavy metals and activated sludge microorganisms, and thus would give a more crucial understanding in the fate of heavy met-als in biological wastewater treatment plants.

1.2 RESULTS AND DISCUSSION

1.2.1 POTENTIOMETRIC TITRATION CURVE AND SURFACE FUNCTIONAL GROUPS OF ACTIVATED SLUDGE

The titration curve of pH vs. NaOH additions is shown in Fig. 1a, and the corresponding derivative of the titration curve is illustrated in Fig. 1b. The activated sludge showed a certain buffering capacity, as indicated by the weak inflection points (Fig. 1a). The derivative of the titration curve (Fig. 1b) gave the equivalence points and the apparent pK_a values for the activated sludge. The peaks indicated a maximum variation in pH corre-sponding to the equivalence points, and the valleys showed a minimum variation in pH, which was an indicator of buffering. Arrows (at pHs of 5.8 and 8.3) represented the corresponding pH values of the titration curve for each equivalence points $(Eq._n)$ [16]. Assuming four binding sites according to the derivative of the titration curve, the pK_a values of the proton-binding sites as well as their contents were estimated using the PROTOFIT 2.1 software, and are shown in Fig. 1c. The pK_a values and their contents of the four binding sites were pK_{a1} 3.57 (0.25 mmol/g), pK_{a2} 5.42 (0.66 mmol/ g), pK_{a3} 7.28 (0.37 mmol/g), and pK_{a4} 9.92 (1.12 mmol/g), which

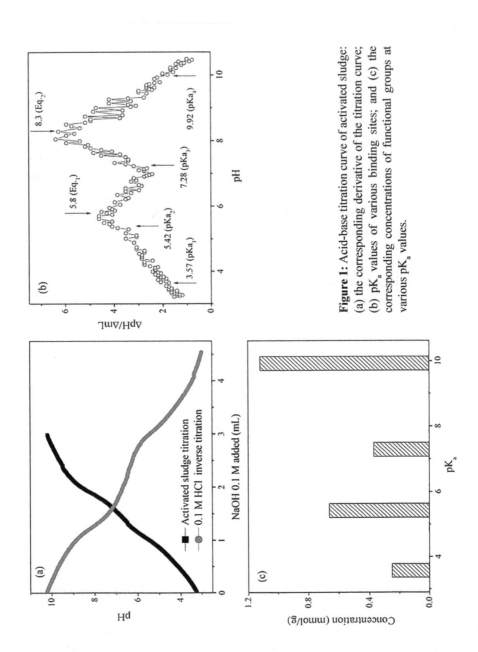

Figure 1: Acid-base titration curve of activated sludge: (a) the corresponding derivative of the titration curve; (b) pK$_a$ values of various binding sites; and (c) the corresponding concentrations of functional groups at various pK$_a$ values.

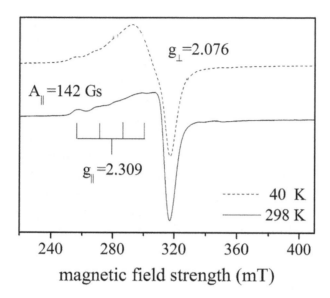

Figure 2: EPR spectra of the Cu(II)- activated sludge complex at 298 K and 40 K.

were assigned as phosphodiester, carboxyl, phosphoryl and hydroxyl/ phenolic groups, respectively [17–19].

1.2.2 EPR RESULTS

The structural characteristics of the Cu(II)-activated sludge complexes were explored using EPR spectroscopy. The EPR spectra displayed an inconspicuous anisotropic copper signal both at 40 K and 298 K (Fig. 2), Theoretically Cu(II) standard should be a typical anisotropic signal with four lines ($2I + 1 = 4$) in parallel region arising from the hyperfine coupling of the $S = 1/2$ electron spin of Cu(II) with its nuclear spin $I = 3/2$. This suggests that the process of cooling had no obvious impact on the structural fate of the Cu(II) complexing, and the structure of Cu(II) complex was different with that of Cu(II) standard. EPR spectra reveal that the Cu(II) center was axial ($g_{\parallel} > g_{\perp} > g_e = 2.00$), with simulated spin Hamiltonian parameters $g_{\parallel} = 2.309$, $g_{\perp} = 2.076$, $A_{\parallel} = 142$ Gs (Fig. 2). The value of hyperfine coupling constant A for Cu(II) was a mean value corresponding to the ^{63}Cu and ^{65}Cu isotopes with their natural abundance. The fact $g_{\parallel} > g_{\perp} > g_e$ indicates that the copper ions had an octahedral coordination with tetragonal distortion of axial elongation (D4h symmetry) due to Jahn-Teller effect [20]. These values were consistent with a d_{x2-y2} ground state for the Cu(II), and the parameter was in the range between those of the cation $Cu(H_2O)_6^{2+}$ ($g_{\parallel} = 2.440$) and those of $Cu(OH)_2$ ($g_{\parallel} = 2.267$). The d_{x2-y2} orbital would increase the destabilization if an axial coordination around Cu(II) ion exists, which causes an increase in the g value from the following modified equations [21]:

$$g_z = g_{\parallel} = g_e + \frac{8|\lambda|}{\Delta E(x^2 - y^2 \rightarrow xy)}$$

$$g_x = g_y = g_{\perp} = g_e + \frac{2|\lambda|}{\Delta E(x^2 - y^2 \rightarrow xz, yz)}$$

where λ represents the spin-orbit coupling constant, ΔE is the difference in the corresponding state energies, and g_e is the g factor of the free electron. Here, $|\lambda| = 830$ cm^{-1} for Cu, $g_e = 2.0023$, thus ΔE can be calculated according to the measured g value, which is used for screening axially symmetric structure in the DFT calculations.

The surface complexation of Cu(II) ion decreased the $g_{\parallel}$-value compared with the value of the free Cu^{2+} ($Cu(H_2O)_6^{2+}$, $g_{\parallel} = 2.440$). The decrease in $g_{\parallel}$ could be associated with the total thermodynamic stability of the formed complex (K_{total}), as shown in eq 3 [21]:

$$\log K_{total} = 84(2.44 - g_{\parallel})$$

where 2.44 is the $g_{\parallel}$ value of the frozen solution aquo ion Cu^{2+}, which has been chosen as a reference point ($\log K_{total} = 0$).

From the relationship between the thermodynamic stability constants of surface complexes and the EPR parameter, the magnitude of Cu(II) complexing strength with the activated sludge could be estimated. The $g_{\parallel}$ value was measured to be 2.309, and thus the surface complex constant (log K_{total}) was estimated to be 11.0, which was comparable to those between Cu and other ligands (e.g., montmorillonite, soil particles, humic substances) with plenty of various functional groups [22], implying the complexing mechanisms between Cu and these ligands might be the same.

1.2.3 XAFS ANALYSIS

The K-edge X-ray absorption spectra of Cu(II) complexing with the activated sludge before and after EPS extraction at various pH values were collected. The Cu K-edge XANES spectra of samples and various standards are shown in Fig. 3. The pre-edge peak at 8976 eV (Arrow 1) was assigned to the 1s to 3d dipole-forbidden electronic transition (probably hybridized by p orbitals of the ligands) [23]. Another peak at 8981 eV (Arrow 2) was resulted from reduced Cu(I) signal. However, the presence of a shoulder at 8981 eV in the Cu absorption edge spectra and the absence of a pre-edge from 8976 eV indicate the presence of Cu(I) in the samples. This peak showed a slightly increase after sequential scans, suggesting a radiation-induced reduction of Cu(II) [24]. The other two peaks in the Cu K-edge XANES spectra (Arrows 3 and 4) were attributed to the 1s to 4p main edge electron transitions. The splitting of the derivative XANES spectra might result from anisotropic square planar symmetry of Cu(II) compounds, or could be referred to the tetragonal distortion of the CuO_6 octahedron due to Jahn-Teller effect. These inflections provided information about the three dimensional geometry and coordination environment of Cu in the Cu complexes. The Peak 3 and Peak 4 corresponded respectively to the 1s $\rightarrow$ 4p and 1s $\rightarrow$ continuum transitions for Cu(II) compounds in octahedral symmetry [25, 26]. The energy gap between the two peaks was 5.0 eV (Fig. 3), attributed to the distortion of $4p_z$ orbit in metalic center. This obtained value was of a similar level to those for other Cu(II) compounds in slightly tetragonally distorted octahedral environments, and was also in accordance with the EPR analytical results. Moreover, the sample spectra in our work had a distinct shoulder peak, which might be due to the degree of axial distortion and the covalence of the equatorial ligands bonded to Cu(II).

After the EPS were extracted from the activated sludge, the XANES spectra of the complexes between Cu and EPS-free activated sludge were shifted to the high energy side because of the increase in oxidation state from the complexes. Also, this phenomenon was observed when pH was increased from 3.0 to 7.0. This might be because the uncomplexed Cu(II) in the equilibrium solution became gradually precipitated to $Cu(OH)_2$, which increased the Cu content in the complex for the XAFS analysis.

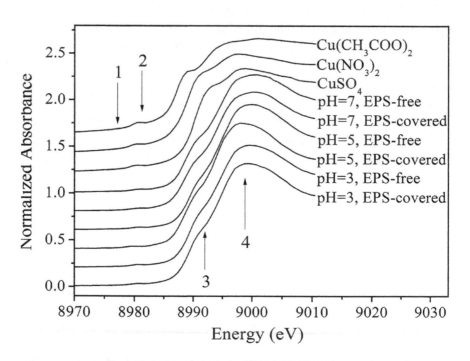

Figure 3: X-ray absorption near-edge structure of reference compounds and Cu- activated sludge under different conditions.

The extraction of EPS influenced the complexation ability of metal ions on the sludge [10], but no significant difference on the binding parameters of the Cu complexes, e.g., bond length and coordination number, was observed. The reason might be that the main functional groups in activated sludge with or without EPS would be similar, although their contents could be changed after the EPS extraction. Thus, the complexing structure between Cu(II) and sludge microorganisms after EPS extraction would not be influenced.

To gain more insight into the molecular structure of the Cu complexing with activated sludge, the EXAFS spectra were fitted according to the standard model of $Cu(NO_3)_2$. The first-shell fit of the EXAFS spectra of the Cu complexes and its corresponding radial structure function (RSF) derived from Fourier transformations, are illustrated in Fig. 4. The position of the peaks in the RSF corresponds to the relative distance (uncorrected for phase shift) between Cu(II) and complexing atoms in local coordination shells. The strongest peak, which appears between 1.44 and 1.50 Å in Fig. 4a, corresponded to the first-shell O atoms. Within the framework of the single scattering approach, the EXAFS spectra fitted well by Levenberg-Marquardt fitting (Fig. 4b), and the results are listed in Table 1. The results show that Cu(II) ions were surrounded by four equatorial oxygen atoms and two axial oxygen atoms. The average $Cu-O_{eq}$ bond length was 1.95 ± 0.01 Å, and the $Cu-O_{ax}$ one was equal to 2.43 ± 0.50 Å. These data are consistent with those obtained from the EPR results and the XANES spectra: distorted octahedra containing four short equatorial bonds and two elongated axial bonds. Also, this agrees with previous studies, in which bond distance for $Cu-O_{eq}$ first shell has been reported to range from 1.92 to 1.97 Å [27–29], while the second shell Cu–O/C ranged from 2.29 to 2.41 Å [30].

As shown in Table 1, the fitting results of the second shell were more susceptible by the extraction of EPS and change of pH than those of the first shell. Because there were usually more coordinated atoms such as C, H and O associated with the complexation of the second shell, which would be influenced by the solution pH and sludge EPS. The bond lengths were in the range of 2.40–2.50 Å and the coordination numbers varied from 1.5 to 1.8 in the second shell. However, the bond lengths were in a range of 1.93–1.95 Å, when the coordination numbers were kept at 4.0 for

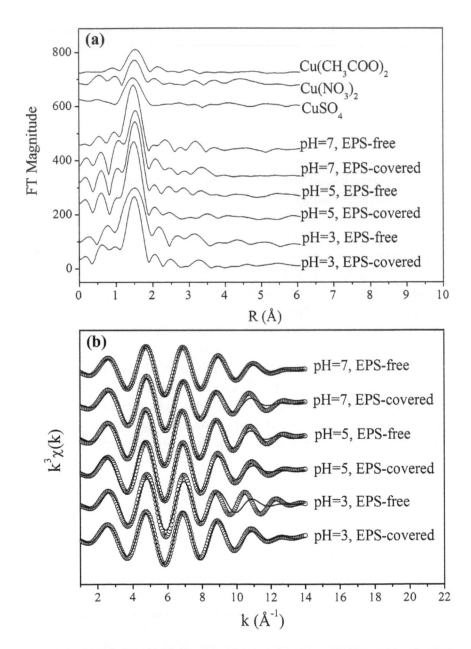

Figure 4: (a) RSF obtained by Fourier transformation of the EXAFS spectrum; and (b) first-shell fit of the EXAFS function of the Cu(II)- activated sludge complex, nonlinear least-squares fits (solid lines) and experimental data (open circles).

the first shell. Those bond lengths were in agreement with the reported values of Cu-hydroxyl or Cu-carboxyl from previous literatures [27–30]. From the microscopic view point, the observed fitting results of Cu(II) K-edge XAFS analysis was not significantly affected after EPS extraction on the bond lengths and coordination numbers. Thus in the short range of a few angstroms surrounded by Cu(II), the bond lengths and coordination numbers were able to stabilize regarding to the complexing structure of Cu and activated sludge microorganisms.

1.2.4 DFT CALCULATION

To further support our interpretation based on EPR and XAFS, the proposed structures of Cu complex were studied by DFT calculations. The Cu(II) center stably existed in a 6-oxygen coordinated octahedron in EPS. The molecular orbital contour plot for the Cu complex and the splitting of energy levels are shown in Fig. S1. Cu 3d orbitals coordinate with π^*, σ orbitals from ligands, and the molecular orbital of Cu complex was split into t_{2g} and e^*_g with the splitting energy ~6000 cm^{-1}, corresponding to the energy $\Delta E = 4071$~5412 cm^{-1} in the EPR analysis (Fig. S1a). The highest occupied molecular orbital (HOMO) shape suggests the low-level orbital of e^*_g of the Cu coordination compound was d_{x2-y2} (Fig. S1b).

Polysaccharides are one of the main compositions for EPS and sludge microorganisms, and the glucose with plenty of hydroxyl groups is one of the main units of polysaccharides [31,32]. The optimized structure of distorted octahedra with Cu center contains four short equatorial bonds and two elongated axial bonds (Fig. 5a), and the angle of distortion is shown in Fig. 5b. The position of octahedrally coordinate around the Cu did not change significantly in the energy minimization. The detailed bond distances and angles of the Cu complexing structure calculated by DFT fitted well with the experimental data (Table 2). The critical bond angles demonstrate that the Cu complex had distorted coordinated octahedral structure, which was comprised of four short equatorial O atoms connected with C from glucose and two elongated axial O with H from H_2O. Four short equatorial O atoms from hydroxyl groups of two glucose molecules form two hexatomic rings with Cu as shown in Fig. 5b. The Cu(II) center is con-

Table 1: Levenberg-Marquardt Fitting of Cu K-edge Bulk XAS Analysis

| | | First shell (Cu-O$_{eq}$) | | | | | Second shell (CU-O$_{ax}$) | |
pH	Cu sample	R (Å)	CN	E$_0$ shift (eV)	σ²(Å²) x 10⁻³	relative error (%)	R (Å)	CN
3.0	EPS-covered	1.93 ± 0.01	4.0 ± 0.3	3.4 ± 0.9	6.9 ± 0.5	5.7	2.40 ± 0.20	1.5 ± 0.9
3.0	EPS-free	1.94 ± 0.01	4.0 ± 0.5	6.3 ± 1.4	7.9 ± 0.8	9.5	2.43 ± 0.50	1.7 ± 0.5
5.0	EPS-covered	1.94 ± 0.01	4.0 ± 0.2	5.8 ± 0.6	6.6 ± 0.4	3.3	2.43 ± 0.10	1.7 ± 0.6
5.0	EPS-free	1.95 ± 0.02	4.0 ± 0.3	5.1 ± 1.0	6.9 ± 0.6	5.3	2.50 ± 0.10	1.8 ± 0.6
7.0	EPS-covered	1.95 ± 0.01	4.0 ± 0.4	7.8 ± 1.1	7.9 ± 0.7	7.3	2.47 ± 0.30	1.7 ± 0.7
7.0	EPS-free	1.94 ± 0.01	4.0 ± 0.3	5.0 ± 0.9	7.6 ± 0.3	2.8	2.41 ± 0.20	1.6 ± 0.7

R: Interatomic distance (Å); CN: Coordination number.
E$_0$ shift: edge energy (eV); s²: Debye-Waller factor (Å²).

nected with two glucose molecules and two H_2O to form hexa-coordinated Cu-sludge complex. The distances of Cu-O5 and Cu-O6 were calculated to be 2.29 Å and 2.28 Å, respectively, which were slightly smaller than those obtained from the XAFS spectra (2.40–2.50 Å) and close to the reported Cu-O$_{ax}$ distance 2.29–2.41 Å. The glucose molecule was bound end-on to the Cu(II) center with Cu–O distances of horizontal plane in a range of 2.05–2.10 Å. This agreed with the experimental Cu-O$_{eq}$ from the XAFS analysis, implying the computational models used for DFT calculation was reasonable. In this case, the Cu complex shown in Fig. 5 was considered to be the most stable one with the lowest energy.

1.2.5 IMPLICATIONS OF THIS WORK

Though the macroscopic interactions between metals and microorganisms have been extensively studied, it is still of great significance to explore such a complexation from a microscopic viewpoint and to observe the microstructure. In the present work, a multi-technique approach was used to probe the complexing characteristics between Cu(II) and microbial cells, and it was proven to be a powerful tool to elucidate such an interaction between heavy metals and activated sludge microorganisms. XAFS and EPR could provide the local structure and geometrical information of Cu. Meanwhile, DFT calculations are able to describe the structural geometry of the Cu complex. Spectroscopic studies confirmed Cu complexation via the formation of inner-sphere surface complexes as the major mechanism of biosorption, and the DFT method was used to calculate the most stable Cu complex structures. These results could be useful for better understanding the speciation, immobilization, transport, and bioavailability of heavy metals in biological wastewater treatment plants, and might also be beneficial for exploring the geochemical cycle of metals in natural waters and soils.

1.3 CONCLUSIONS

The microscopic-level complexing between heavy metal Cu(II) and activated sludge has been studied. Spectroscopic results indicate that Cu complexed with activated sludge via the formation of inner-sphere surface complexes of octahedral coordination with tetragonal distortion of axial elongation, containing four short equatorial bonds with a mean $Cu\text{-}O_{eq}$ bond distance of 1.95 Å and two elongated axial bonds with a $Cu\text{-}O_{ax}$ bond distance of 2.43 Å. DFT calculation was able to explain the structural geometry of the Cu complex, and fitted well with the experimental results. These results would be useful for understanding the mechanism of the fate and species of Cu(II) in biological wastewater treatment plants.

1.4 METHODS

1.4.1 EPS EXTRACTION FROM ACTIVATED SLUDGE AND CU(II) ADSORPTION TESTS

The activated sludge was harvested from a laboratory-scale sequencing batch reactor fed with a synthetic wastewater, which the composition was: CH_3COONa, 640 mg L^{-1}; NH_4Cl, 95 mg L^{-1}; KH_2PO_4, 22.5 mg L^{-1}; $CaCl_2$, 11.5 mg L^{-1}; $MgSO_4$, 12 mg L^{-1} and 10 mL of trace element solution. The C: N: P ratio was 100:5:1. The reactor was operated sequentially in 6 h cycles, with 7 min of substrate filling, 342 min of aeration, 2 min of settling and 3 min of effluent withdraw. The sludge concentration was 3.6 g/L, and the ratio of volatile suspended solids (VSS) to suspended solids (SS) was 0.82, implying the high microorganism contents in the sludge. Before experiments, the sludge microorganisms were washed by distilled water for three times to remove the residual substrate and metabolic products. The EPS were extracted using the cation exchange resin (CER) technique as described previously [33]. Briefly, the activated sludge sample, washed twice with 50 mM NaCl solution, was stirred for 12 h at 200 rpm and 4°C after the CER addition (60 g/g SS). Subsequently, the solutions were centrifuged to remove CER and remaining sludge components. The superna-

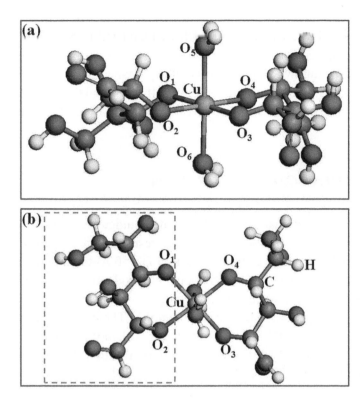

Figure 5: DFT optimized results: (a) Local structures of Cu complex; and (b) Top view to show the bond angle.

tants were then filtered through 0.45 μm cellulose acetate membranes and used as the EPS fraction.

The process of Cu(II) biosorption was conducted using 50 mL shaking flasks. The detailed information about characterization of the Cu(II) biosorption onto sludge has been given in the Supplementary Information. The maximum Cu retention capacity of the activated sludge at pH 5.0 was calculated to be 25.7 mg/g through Langmuir adsorption isotherm model (Fig. S2). The microstructure of Cu(II) complexing with the functional groups in activated sludge were explored using XAFS and EPR techniques, and high content of complexed Cu(II) on sludge would be beneficial for the accuracy of XAFS and EPR analysis. Herein, Cu(II) concentration at 64 mg/L and 2.08 g of activated sludge were added to the flasks and mixed. The ionic strength was adjusted to 50 mM NaCl. The pH was adjusted to 3.0, 5.0 and 7.0. After 12 h of equilibrium, the sludge after complexing with Cu(II) was carefully washed twice by 50 mM NaCl solution to remove the remaining free Cu(II) ions, and then used for XAFS and EPR analysis. The sludge after EPS extraction was also used for comparison.

1.4.2 POTENTIOMETRIC TITRATION

The surface functional groups of activated sludge were determined with the potentiometric titration. The sludge resuspended in 40 mL solution (6.0 g-dry weight/L) was titrated using a DL 50 Automatic titrator (Mettler Toledo Co., Switzerland) with a pH electrode of 0.001 precision. The titration was conducted under nitrogen gas conditions at 25°C. The solution ionic strength was adjusted to 0.01 mol/L. The initial pH was adjusted to 3.0 by 1 mol/L HCl, and was titrated by 0.1 mol/L NaOH with 10 μL increments until pH 11.0 was reached. The titration data obtained were analyzed using the PROTOFIT 2.1 software [34].

1.4.3 XAFS MEASUREMENTS AND ANALYSIS

The Cu K-edge XAFS spectra of Cu(II) sorbed on the activated sludge were measured at the U7C beamline of the National Synchrotron Radiation Laboratory (NSRL), Hefei, China. The XAFS signals were collected in a fluorescence mode with a seven-element high-purity Ge solid detector. The electron beam energy was 0.8 GeV and the maximum stored current was 300 mA. A double crystal Si (1 1 1) monochromator was used. Energy calibration was monitored using a Cu metallic foil with the first inflection of the absorption edge set to 8979 eV. The spectra were recorded in the energy range of 8779–9774 eV covering the copper K- edge (<8979 eV) with intervals of 0.5 eV for XANES and 2 eV for EXAFS. An integration time of 5.0 s per point was used in both cases. The obtained data of all the standards and samples in three scans were averaged to improve the signal to noise ratios (S/N), which was more than 103 under the experimental conditions. The raw data analysis was performed by using the NSRL-XAFS 3.0 software package according to the standard data analysis procedures [35].

1.4.4 EPR MEASUREMENT

The EPR spectra of Cu(II) complex with activated sludge before EPS extraction at pH 5 5.0 were measured using an EMX spectrometer (Bruker Co., Germany) at 40 K and 298 K, which was operated at X-band frequency (9.72 GHz) with a 100 kHz modulation frequency. Sample pretreatments for the EPR measurement were consistent with those for the XAFS measurement. All measurements were repeated four times, and a gradient cooling method was applied: a) solid ethanol (190 K); b) liquid nitrogen (77 K); and c) liquid helium (40 K), to avoid the breaking of the EPR sample tubes. The EPR spectra were recorded at a microwave power of 2 mW and modulation amplitude of 5.0 G.

1.4.5 DFT CALCULATION

All the calculations were completed using the DMol^3 module [36,37] of the Materials Studio Program. The minimum-energy geometry structures of the Cu complex were determined by an all-electron method within the Perdew-Wang 91 (PW91) form of generalized gradient approximation (GGA) [38,39] for the exchange-correlation term. The double precision numerical basis sets including p polarization (DNP) were adopted. A spin-polarized scheme was employed to deal with the open-shell systems. The energy in each geometry optimization cycle was converged to within 2 x 10^{-5} Hartree with a maximum displacement and force of 5 x 10^{-3} Å and 4 x 10^{-3} Hartree/Å, respectively.

The computational models of the active sites were constructed from carboxyl, hydroxyl and glucose units, which were one of the main compositions for EPS and sludge microorganisms [31]. Combined with XAFS and EPR results, geometry optimization of Cu complex was performed with frequent updates of the force constants in order to stay as close as possible to the lowest energy pathway. The complex was regarded to reach a most stable structure when the energy came to the lowest horizontal line. Furthermore, the splitting result of orbital energy level and the analysis of orbital elements could also be obtained.

REFERENCES

1. Vaiopoulou, E. & Gikas, P. Effects of chromium on activated sludge and on the performance of wastewater treatment plants: A review. Water Res. 46, 549–570 (2012).
2. Park, C. & Novak, J. T. Characterization of activated sludge exocellular polymers using several cation-associated extraction methods. Water Res. 41, 1679–1688 (2007).
3. Tong, M., Long, G., Jiang, X. & Kim, H. N. Contribution of extracellular polymeric substances on representative Gram Negative and Gram Positive bacterial deposition in porous media. Environ. Sci. Technol. 44, 2393–2399 (2010).
4. Das, S. K., Ghosh, P., Ghosh, I. & Guha, A. K. Adsorption of rhodamine B on Rhizopus oryzae: Role of functional groups and cell wall components. Colloid Surf. B. 65, 30–34 (2008).
5. Ueshima, M. et al. Cd adsorption onto *Pseudomonas putida* in the presence and absence of extracellular polymeric substances. Geochim. Cosmochim. Acta. 72, 5885–5895 (2008).
6. Fang, L. C. et al. Role of extracellular polymeric substances in Cu(II) adsorption on *Bacillus subtilis* and *Pseudomonas putida*. Bioresour. Technol. 102, 1137–1141 (2011).
7. Benaïssa, H. & Elouchdi, M. A. Biosorption of copper (II) ions from synthetic aqueous solutions by drying bed activated sludge. J. Hazard. Mater. 194, 69–78 (2011).
8. Laurent, J., Casellas, M., Carrère, H. & Dagot, C. Effects of thermal hydrolysis on activated sludge solubilization, surface properties and heavy metals biosorption. Chem. Eng. J. 166, 841–849 (2011).
9. Sag, Y., Tatar, B. & Kutsal, T. Biosorption of Pb(II) and Cu(II) by activated sludge in batch and continuous-flow stirred reactors. Bioresour. Technol. 87, 27–33 (2003).
10. Ha, J., Gelabert, A., Spormann, A. M. & Brown, G. E. Role of extracellular polymeric substances in metal ion complexation on *Shewanella oneidensis*: Batch uptake, thermodynamic modeling, ATR-FTIR and EXAFS study. Geochim. Cosmochim. Acta. 74, 1–15 (2010).
11. Drzewiecka, A. et al. Synthesis and structural studies of novel Cu(II) complexes with hydroxy derivatives of benzo[b]furan and coumarin. Polyhedron 43, 71–80 (2012).
12. Wang, X. Y., Deng, X. T. & Wang, C. G. Bis (acetato-k²O,O′) diaquacopper(II). Acta Crystallogr. Sect. E.-Struct Rep. 62, M3578–M3579 (2006).
13. Adamescu, A., Hamilton, I. P. & Al-Abadleh, H. A. Thermodynamics of dimethylarsinic acid and arsenate interactions with hydrated iron- (oxyhydr)oxide clusters: DFT calculations. Environ. Sci. Technol. 45, 10438–10444 (2011).

14. 1He, G. Z., Pan, G., Zhang, M. Y. & Waychunas, G. A. Coordination structure of adsorbed Zn(II) at water-TiO$_2$ interfaces. Environ. Sci. Technol. 45, 1873 1879 (2011).
15. Jin, X., Yan, Y., Shi, W. & Bi, S. Density functional theory studies on the structures and water-exchange reactions of aqueous Al(III)-oxalate complexes. Environ. Sci. Technol. 45, 10082–10090 (2011).
16. Braissant, O. et al. Exopolymeric substances of sulfate-reducing bacteria: Interactions with calcium at alkaline pH and implication for formation of carbonate minerals. Geobiol. 5, 401–411 (2007).
17. Laurent, J., Pierra, M., Casellas, M. & Dagot, C. Fate of cadmium in activated sludge after changing its physico-chemical properties by thermal treatment. Chemosphere. 77, 771–777 (2009).
18. Lee, S. M. & Davis, A. P. Removal of Cu(II) and Cd(II) from aqueous solution by seafood processing waste sludge. Water Res. 35, 534–540 (2001).
19. Liu, H. & Fang, H. H. P. Characterization of electrostatic binding sites of extracellular polymers by linear programming analysis of titration data. Biotechnol. Bioeng. 80, 806–811 (2002).
20. Bahranowski, K., Dula, R., Labanowska, M. & Serwicka, E. M. ESR study of Cu centers supported on Al-, Ti-, and Zr-pillared montmorillonite clays. Appl. Spectrosc. 50, 1439–1445 (1996).
21. Motschi, H. Correlation of EPR-parameters with thermodynamic stability constants for copper(II) complexes - Cu(II) electron paramagnetic resonance as a probe for the surface complexation at the water/oxide interface. Colloids Surf. 9, 333–347 (1984).
22. Flogeac, K., Guillon, E. & Aplincourt, M. Surface complexation of copper(II) on soil particles: EPR and XAFS studies. Environ. Sci. Technol. 38, 3098–3103 (2004).
23. Frenkel, A. I., Korshin, G. V. & Ankudinov, A. L. XANES study of Cu^{2+}- binding sites in aquatic humic substances. Environ. Sci. Technol. 34, 2138–2142 (2000).
24. Strawn, D. G. & Baker, L. L. Speciation of Cu in a contaminated agricultural soil measured by XAFS, μ-XAFS and μ-XRF. Environ. Sci. Technol. 42, 37–42 (2008).
25. Furnare, L. J., Vailionis, A. & Strawn, D. G. Polarized XANES and EXAFS spectroscopic investigation into copper(II) complexes on vermiculite. Geochim. Cosmochim. Acta. 69, 5219–5231 (2005).
26. Choy, J. H., Yoon, J. B. & Jung, H. Polarization-dependent X-ray absorption spectroscopic study of [Cu(cyclam)](21)-intercalated saponite. J. Phys. Chem. B. 106, 11120–11126 (2002).
27. Karlsson, T., Persson, P. & Skyllberg, U. Complexation of copper(II) in organic soils and in dissolved organic matter - EXAFS evidence for chelate ring structures. Environ. Sci. Technol. 40, 2623–2628 (2006).
28. Hsiao, M. C., Wang, H. P. & Yang, Y. W. EXAFS and XANES studies of copper in a solidified fly ash. Environ. Sci. Technol 35, 2532–2535 (2001).
29. Korshin, G. V., Frenkel, A. I. & Stern, E. A. EXAFS study of the inner shell structure in copper(II) complexes with humic substances. Environ. Sci. Technol. 32, 2699–2705 (1998).
30. Scheinost, A. C., Abend, S., Pandya, K. I. & Sparks, D. L. Kinetic controls on Cu and Pb sorption by ferrihydrite. Environ. Sci. Technol. 35, 1090–1096 (2001).

31. Wang, Z., Choi, O. & Seo, Y. Relative contribution of biomolecules in bacterial extracellular polymeric substances to disinfection byproduct formation. Environ. Sci. Technol. 47, 9764–9773 (2013).
32. Wang, Z., Choi, O. & Seo, Y. Relative contribution of biomolecules in bacterial extracellular polymeric substances to disinfection byproduct formation. Environ. Sci. Technol. 47, 9764–9773 (2013).
33. Dignac, M. F. et al. Fate of wastewater organic pollution during activated sludge treatment: nature of residual organic matter. Water Res. 34, 4185–4194 (2000).
34. Sheng, G. P., Zhang, M. L. & Yu, H. Q. Characterization of adsorption properties of extracellular polymeric substances (EPS) extracted from sludge. Colloids Surf. B-Biointerfaces 62, 83–90 (2008).
35. Sheng, G. P., Zhang, M. L. & Yu, H. Q. Characterization of adsorption properties of extracellular polymeric substances (EPS) extracted from sludge. Colloids Surf. B-Biointerfaces 62, 83–90 (2008).
36. Turner, B. F. & Fein, J. B. Protofit: A program for determining surface protonation constants from titration data. Comput. Geosci. 32, 1344–1356 (2006).
37. Sayers, D. E. & Bunker, B. A. Data analysis. X-ray Absorption, Principles, Applications, Techniques of EXAFS, SEXAFS and XANES [Koningsberger, D.C., Prins, R. (eds.)] [211–253] (Wiley, New York, 1998).
38. Delley, B. Fast calculation of electrostatics in crystals and large molecules. J. Phys. Chem. 100, 6107–6110 (1996).
39. Delley, B. From molecules to solids with the DMol(3) approach. J. Phys. Chem. 113, 7756–7764 (2000).
40. Perdew, J. P. et al. Atoms, molecules, solids and surfaces: Applications of the generalized gradient approximation for exchange and correlation. Phys. Rev. B. 46, 6671–6687 (1992).
41. Perdew, J. P. & Wang, Y. Accurate and simple analytic representation of the electron-gas correlation energy. Phys. Rev. B. Condens. Matter. 45, 13244–13249 (1992).

There are several supplemental files that are not available in this version of the article. To view this additional information, please use the citation on the first page of this chapter.

CHAPTER 2

A Novel Protocol for Model Calibration in Biological Wastewater Treatment

AO ZHU, JIANHUA GUO, BING-JIE NI, SHUYING WANG, QING YANG, AND YONGZHEN PENG

2.1 INTRODUCTION

Activated sludge is the most widely used biological technology for treating domestic and industrial wastewater. After its development with 100 years of history, many novel and modified processes have been developed to meet the more and more stringent discharge and emission limits. However, most of operating systems are suffering some drawbacks, such as substantial energy consumption, excessive greenhouse gas emission, and labour-intensive industry. As a powerful tool, Activated Sludge Models (ASMs) have proven to be very useful in process design, operation and optimization [1,2]. To date, ASMs for the simulation of biological nutrients removal processes have been updated from the first version of ASM1 to more compli-

A Novel Protocol for Model Calibration in Biological Wastewater Treatment. © *Zhu A, Guo J, Ni B-J, Wang S, Yang Q, and Peng Y.* Scientific Reports *5,3493 (2015). doi:10.1038.srep08493. Licensed under a Creative Commons Attribution 4.0 International License, http://creativecommons.org/licenses/by/4.0/.*

cated extensions, including ASM2, ASM2d, and ASM3 [3], and further to the extended ASM3s [4–10] in order to satisfy various requirements.

However, ASMs are large and overparameterized models in terms of having many stoichiometric and kinetic parameters. Some of the model parameters as well as the model structure have to be adjusted, since microbial community structure and dominant species can vary in different wastewater treatment systems with different influent characteristics or operation schemes [6,7,11–14]. In addition, the collected data from full-scale plants as well as pilot- or lab-scale reactors can hardly provide reliable estimations of all the parameters simultaneously due to the well-known problem of poorly identifiable parameters [15–20]. Thus, the approach to properly select the subsets of parameters for model calibration plays a crucial role on simulation results and model applications [21–25].

Until now, substantial studies have been conducted to develop effective model calibration approaches, which can be distinguished into two major categories: the conventional experience-based approaches and the systems analysis approaches [22]. The experience-based approaches that were proposed by WERF, BIOMATH, STOWA, CALAGUA, HSG etc., import programmatic flow based on experts' knowledge and experience [22,26,27]. However, these calibration protocols require specific experimental designs and data processing methods to decouple the ASMs to be small and simple sub ones. The data for model calibration and validation may be from different feeding and operation conditions, and some of the parameters are fixed in order to obtain accurate estimation of the concerning ones. This approach might also ignore shifts in microbial community, eventually resulting in error estimations. Recently, the Good Modeling Practice (GMP) task group from IWA have emphasized that a standardized modeling procedure is needed to distinguish parameter subset that is identifiable with the available data for experience-based approaches [28].

Systematical analysis approaches have attracted much more attention [15,18,29–32], which mainly consist of parameter identification, sensitivity analysis, and error propagation [32]. Conventional parameter subset identifiability analyzing methods are mainly based on the local sensitivity analysis [21,22]. Recently, global sensitivity analysis (GSA) becomes a promising method for parameter identifiability analysis [33–35], such as the Morris screening method [36], Fourier Amplitude Sensitivity Testing

(FAST), Extended-FAST, the Sobol's method [37–41], and standardized regression coefficient (SRC) method [33]. Structural identification through Taylor series expansion [16], generating series [42], similarity transformation approach [43] and differential algebra approach [17,44], have been proposed in order to screen out the identifiable combinations. Based on the symbolic algebra system, structural identification has been applied to ASM-type simple sub-models successfully [16,17]. However, structural identification based on symbolic algebra system could be time consuming and unreliable due to the large number of parameters in ASMs. Therefore, it is highly desired to develop a numerical integrated approach by including both proper global parameter sensitivity analysis method and correlation analysis method for the identifiability of the parameters of the ASMs.

The main objective of this study is to develop a novel systematical approach, namely Numerical Optimal Approaching Procedure (NOAP), for efficient calibration and validation of ASMs, taking an extended ASM3 as example. The NOAP integrates a new screening global parameter sensitivity analysis method (Derivative based Global Sensitivity Measures, DGSM) and a numerical correlation analysis method (based on the pseudo-global covariance matrix), with Genetic Algorithm (GA) as the global optimization (GO) method for parameter estimation. The validity and applicability of the approach for efficient model calibration are tested by independent experimental data from a Sequential Batch Reactor (SBR) activated sludge system [45] and a Continuous Stirred-Tank Reactor (CSTR) activated sludge system [46].

2.2 RESULTS

2.2.1 DEVELOPMENT OF THE NOAP APPROACH

The iterative algorithm of NOAP approach, as shown in the main scheme flow in Fig. 1, mainly consists of four steps in succession.

Step I: Model selection and data preparation. The user should firstly choose an appropriate model and extend it to satisfy the specific requirement from the current available models in previous studies [1,47]. Data collected from lab-, pilot-, and full-scale reactors are often used, but the amount and quality need reconciliation for the basic request of model calibration. The data collection and reconciliation can refer to the procedures suggested by Rieger et al. [28].

Step II: GSA for factors fixing. Global parameter sensitivity analysis would be performed only if (i) the initial values of the state variables are set in a reasonable range based on data analysis and ii) parameter boundaries (lowest and highest) for global sensitivity analysis are proper. A fitting goodness criterion (such as A = 85%) between data and simulation results is used to judge parameter subset remained after factor fixing from DGSM analysis result. If the result is satisfied $(R \geq A)$, the parameters with lower sensitivity will be removed with trial, then repeating estimation trial until a proper small enough parameter subset is selected out. Otherwise, more non-sensitive ones should be added into the target subset one by one with the iterative parameter estimation trial. If parameters are all estimated simultaneously without a satisfied result, boundaries of the parameters or model structure might be improper and need further modifications.

Step III: For the scale reduced parameter subset, a pseudo-global correlation matrix is calculated. If the correlation coefficient of any parameter pair is high enough (>0.95 as example), then such parameter combinations are located to be highly correlated parameters due to the statistical average effect of the numerical algorithm. Afterwards, less interested members and correlation crossing ones (high correlated with not only just one parameter) can be fixed.

Step IV: The final parameter estimation is performed to check the efficiency of the procedure. Another fitting goodness criterion (such as B = 95%) is set to roughly test the quality of the observables' collected data.

2.2.2 PRELIMINARY MODEL CALIBRATION BASED ON CONVENTIONAL APPROACH

According to Kaelin et al. [6], the parameter subset [μ_{AOB}, μ_{NOB}, $\eta_{H,NO3}$, $\eta_{H,NO2}$, $\eta_{H'end.NO3}$, $\eta_{H,end.NO2}$] is firstly adopted and used to perform preliminary model calibration. Through this step, it is expected to: (1) determine whether the experimental data from the target system could be simulated by the proposed model; (2) suggest proper initial values of the state vari-

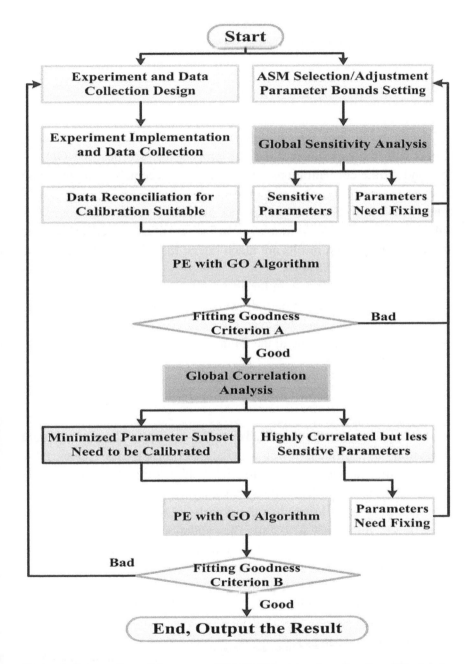

Figure 1: Scheme flow of the proposed NOAP (PE, parameter estimation; GO, global optimization).

ables; and (3) verify directly whether the reported parameter subset from literature is efficient or not.

2.2.2.1 SCENARIO 1: SBR SYSTEM

As shown in Fig. 2(a), the simulated results (Sim-Re series) have a good fitness with the experimental data in terms of the profiles of COD, nitrite and nitrate concentrations. However, an unacceptable accumulation of ammonium during anoxic period is observed in the model simulations, which might be due to improper parameter values. Moreover, the validation (Sim-Re series) illustrated in Fig. 2(b) further verifies that the simulated ammonium results are much higher than the experimental data ($R^2(S_{NH}$ Sim-Re) = 0.5011).

Regarding the SBR system for nitrogen removal, the parameter subset selected from literature can partially calibrate the model to describe most of the experimental data. However, the extra efforts are needed to select other parameter subsets due to the failure of model validation.

2.2.2.2 SCENARIO 2: CSTR SYSTEM

As shown in Fig. 3(a), the simulated (Sim- Re series) COD and ammonium concentrations in effluent fit well with the experimental data. However, the simulated nitrite concentrations are slightly higher than the experimental data (R^2 (Eff. S_{NO2} Sim-Re) = 20.3617), and the simulated nitrate concentrations are distinctly lower than the experimental data after the 30th day (R^2 (Eff. S_{NO3} Sim-Re) = 21.0290). The validation results (Sim-Re series) in Fig. 3(b) suggest that all the measured nitrogen species concentrations (including ammonium, nitrite and nitrate) deviate distinctly from the experiment data (R^2 (Eff. S_{NH} Sim-Re) = 23.2151; R^2 (Eff. S_{NO2} Sim-Re) = 216.4224; R^2 (Eff. S_{NO3} Sim-Re) = 23.3828). This might be attributed to the improper parameter values of the dissolved oxygen (DO)-related kinetic coefficients, such as $K_{H, O2, inh}$, $K_{AOB, O2,}$ and $K_{NOB, O2}$. Hence, the model calibration is failed when applying the literature reported parameter subset [6] into the CSTR system.

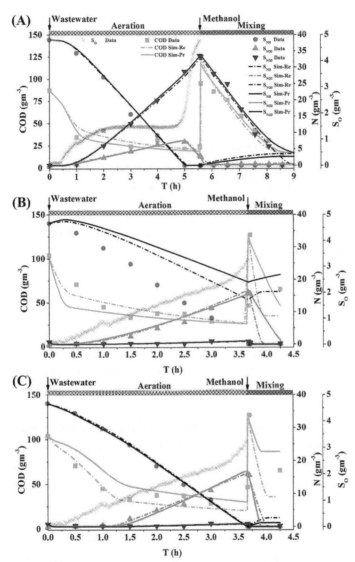

Figure 2: Calibration and validation results with the parameter subset suggested by the reference 6 and the proposed procedure NOAP for target SBR system. (A, Calibration results with the parameter subset recommended by the reference (Sim-Re) [6] and the NOAP procedure (Sim-Pr); B, Validation for the calibration results as Fig. 2(a) presented, legends are the same as Fig. 2(a); C, Validation adjusted for the calibration with the parameter subset recommended by the reference [6] and the proposed procedure NOAP with the suggestions of the global parameter sensitivity and correlation analysis results (Here gH,NO$_2$ is added to the calibration subset), legends are the same as Fig. 2(a).)

The preliminary calibration results from SBR and CSTR systems indicate that the parameter sensitivity vary across different activated sludge systems in the application of same model. Moreover, the parameter subset reported in literature could lead to poor model calibration results if fixing the remained parameters (e.g., the DO related parameters in Scenario 2). Thus, reliable parameter sensitivity analysis should be imported for model application in different activated sludge systems. It should be noted that the parameter subset reported in literature could be still useful for qualitatively model predictions, although it may not serve as a precise and quantitative predictor to match all experiment data.

2.2.3 PARAMETER SENSITIVITY AND CORRELATION ANALYSIS

According to the preliminary calibration and validation, it can be concluded that the extended ASM3 model could predict the data only if a proper parameter subset is selected. However, the parameter subset reported in literature [6] cannot achieve satisfied calibration or validation results for both SBR and CSTR systems. Thus, numeric global parameter sensitivity and correlation analysis are performed to determine a more proper parameter subset.

In order to assess the identification of all the related parameters listed in Table S3 (SI), parameter sensitivity and correlation analysis are conducted for both SBR and CSTR systems. The proposed iterative trials (Fig. 1) are implemented in order to capture the best-fit parameter values. The results are shown in Fig. 4 and Fig. 5, respectively. It can be found that both sensitivity analysis and correlation analysis results are quite different between the SBR system and CSTR system. Furthermore, parameters indexed from 1 to 6 in Table S4 (SI) exhibit a similar low sensitivity for both SBR and CSTR systems, suggesting the composition coefficients could be fixed as default values due to their low sensitivity. However, most of the composition coefficients strongly depend on the influent wastewater characteristics. Therefore, it is recommended that the N content of biomass can be fixed as default values from references, while others (e.g. COD transformation fractions) need to be obtained experimentally.

For the SBR system, parameters including $Y_{H,NO2}$, $Y_{STO,O2}$, $Y_{STO,NO3}$, $Y_{STO,NO2}$, Y_{AOB}, k_{STO}, μ_{AOB}, μ_{NOB}, $\eta_{H,NO2}$ and $K_{AOB,O2}$ show high sensitivity, with their normalized relative influences reach 0.01 or even higher (Fig. 4(a)). Meanwhile, according to Fig. 5(a), the highly correlated parameter pairs are $Y_{H,NO2}$ vs. $Y_{STO,NO2}$ ($R^2 = 0.98$), Y_{AOB} vs. μ_{NOB} ($R^2 = 0.99$), Y_{AOB} vs. $K_{AOB,O2}$ ($R^2 = 0.99$), and μ_{AOB} vs. $K_{AOB,O2}$ ($R^2 = 0.99$). In addition, the sensitivity of $Y_{STO,NO2}$ is much higher than $Y_{H,NO2}$ and Y_{AOB}, and $K_{AOB,O2}$ is highly correlated with other parameters, thus, $Y_{H,NO2}$, Y_{AOB} and $K_{AOB,O2}$ should be eliminated from the parameter subset. Finally, a parameter subset [$Y_{STO,O2}$, $Y_{STO,NO3}$, $Y_{STO,NO2}$, k_{STO}, μ_{AOB}, μ_{NOB}, $\eta_{H,NO2}$] has been selected for the SBR system. Similarly, the parameter subset selected for CSTR system has been determined as [$Y_{STO,NO3}$, $Y_{STO,NO2}$, $\eta_{H,NO3}$, $\eta_{H,NO2}$, $K_{H,O2.inh}$, $K_{H,NO3}$, $K_{AOB,O2}$, $K_{NOB,O2}$]. The corresponding calibration and validation achieve satisfied results as shown in Fig. 2(b) and Fig. 3(b).

Optimal parameter subsets of SBR and CSTR systems for the calibration of the extended ASM3 are highly different. In the SBR system, three processes dominate the biological activities, which are storage of SS by heterotrophic organisms, growth of autotrophic organisms and denitrification via nitrite as the electron acceptor. These results are consistent with the experimental observations in the target SBR system, in which simultaneous nitrification and denitrification (SND) was observed and the nitrogen loss due to SND under aerobic condition was about 10 ~ 20 mg/L [45]. In contrast, for the CSTR system, the dominated biological activities include storage of SS by heterotrophic organisms, denitrification and inhibition of DO on heterotrophic denitrification and autotrophic growth. Hence, different optimal parameter subsets of SBR and CSTR systems may be mainly attributed to the different operation conditions, which lead to different dominant biological activities. From the modeling perspective, different initial values of the state variables, inputs like DO, influent COD and ammonium and the parameter boundaries would lead to different optimal parameter subsets.

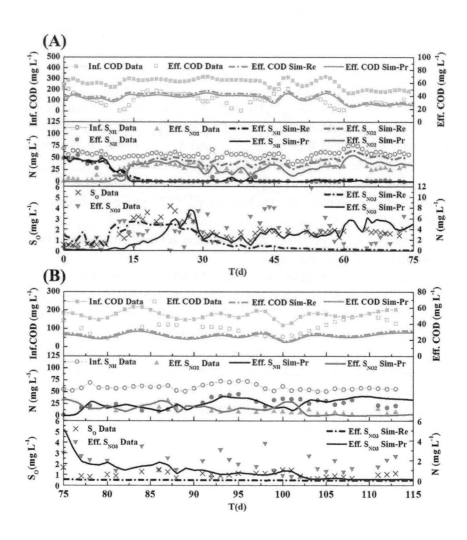

Figure 3: Calibration and validation with the parameter subset recommended by the reference [6] and the proposed procedure NOAP for target CSTR system (A, Calibration results with the parameter subset suggested by the reference [6] and the NOAP procedure (Sim-Pr); B, Validation for the calibration results as Fig. 3(a) presented.)

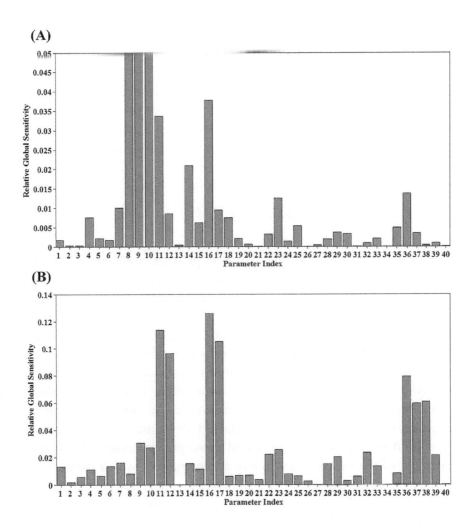

Figure 4: Global parameter sensitivity analysis results. (A, Parameter sensitivity analysis for the SBR system (parameter 8, 9 and 10 reached 0.20, 0.13 and 0.46, respectively; B, Parameter sensitivity analysis results for the CSTR system, The indexes and their corresponding parameters are listed as the follows: 1- $i_{N,SS}$, 2- $i_{N,XI}$, 3- $i_{N,BM}$, 4- f_{XI}, 5-$Y_{H,O2}$, 6-$Y_{H,NO3}$, 7-$Y_{H,NO2}$, 8-$Y_{STO,O2}$, 9-$Y_{STO,NO3}$, 10-$Y_{STO,NO2}$, 11-Y_{AOB}, 12-Y_{NOB}, 13-k_{H}, 14- k_{STO}, 15-μ_{H}, 16- μ_{AOB}, 17-μ_{NOB}, 18-$b_{H,O2}$, 19-$b_{STO,O2}$, 20-b_{AOB}, 21-b_{NOB}, 22-$g_{H,NO3}$, 23-$g_{H,NO2}$, 24-$g_{H,end:NO3}$, 25-$g_{H,end:NO2}$, 26-$g_{N,end}$, 27-K_{X}, 28-$K_{H,O2}$, 29-$K_{H,O2:inh}$, 30- $K_{H,SS}$, 31-$K_{H,NH4}$, 32-$K_{H,NO3}$, 33-$K_{H,NO2}$, 34-$K_{H,AL}$, 35-$K_{H,STO}$, 36-$K_{AOB,O2}$, 37-$K_{NOB,O2}$, 38-$K_{AOB,NH4}$, 39-$K_{NOB,NO2}$, 40-$K_{N,ALK}$. The meanings of each parameter can be found in Table S4, SI).

2.2.4 PARAMETER ESTIMATION AND MODEL VALIDATION BASED ON THE NOAP APPROACH

In order to compare with the results obtained through the experience approach (i.e., preliminary calibration), the parameter subsets for SBR and CSTR systems determined by NOAP are further utilized for calibration and validation in this section. All the parameter estimation results can be found in Table 1.

2.2.4.1 SCENARIO 1: SBR SYSTEM

As illustrated in Fig. 2(a), the subset of parameters determined by the NOAP can efficiently calibrate the extended ASM3 through the GO algorithm (see Sim-Pr series). The results indicate that a satisfied calibration could be obtained by different parameter subsets under the condition of limited data availability. Compared to the results of the traditional experience-based approach's calibration results (Fig. 2(a) Sim-Re series), the accumulation of ammonium during anoxic period in the NOAP calibration (Fig. 2(a) Sim-Pr series) is less, demonstrating a better fitting ability of the proposed NOAP approach. Furthermore, the nitrate reduction rate becomes slower when the biodegradable COD concentration is insufficient for denitrification. The simulated nitrate profile deviate slightly from the trend of the experimental nitrate concentrations, which indicates the parameter subset determined by the NOAP is more sensitive than that suggested by the traditional experience-based approach.

Although the calibration is successfully achieved, the initial validation (Fig. 2(b) Sim-Pr series) is still failed ($R^2(S_{NH}$ Sim-Pr) = 0.0997), similar with the validation by the experience-based parameter subset (Fig. 2(b) Sim-Re series). However, the proposed NOAP provides possible solutions to improve the simulation performance. After the iterative parameter subset selection procedure by NOAP, the parameter $K_{H,O2:inh}$ is further added into the parameter subset for re-calibration of the model. As shown in Fig. 2 (c), the performance of re-calibration for both parameter subsets (suggested by the traditional experience-based approach and the NOAP, respectively) are improved after adding the parameter $K_{H,O2:inh}$ in the pa-

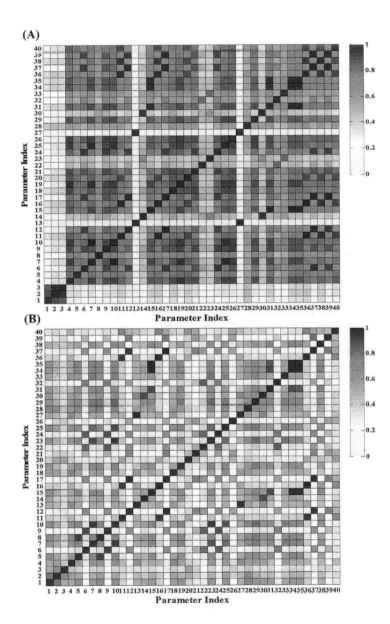

Figure 5: Global parameter correlation analysis results (color of the off-diagonal elements represents related parameters' correlation, between -1 and 1) (A, Parameter correlation analysis result matrix for the SBR system; B, Parameter correlation analysis result matrix for the CSTR system. The indexes and their corresponding parameters are the same with Fig. 4.)

rameter subset. In addition, the parameter subset $[Y_{STO,O2},\ Y_{STO,NO3},\ Y_{STO,NO2},\ k_{STO},\ \mu_{AOB},\ \mu_{NOB},\ K_{H,O2:inh},\ \eta_{H,NO2}]$ determined by the NOAP can reach a better fitting goodness, as the simulated ammonium concentrations match better with the original data during the anoxic period (Fig. 2 (c)). Thus, the proposed NOAP demonstrates its ability to provide additional information for improving model calibration.

2.2.4.2 SCENARIO 2: CSTR SYSTEM

As present as Fig. 3, the calibration results (see Sim-Pr series) with parameter subset determined by the NOAP procedure achieve a much better fitting goodness with the experimental data (as shown in Fig. 3(a)). The validation of the calibrated parameters (Fig. 3(b)) also shows a good fitting except the period from day 99 to day 101, in which a lower ammonium and higher nitrite concentrations are predicted. The reason for such phenomena might result from the use of DO as an input for simulation and the DO concentrations during the period are higher than the real concentrations. It can be concluded that parameter subset $[Y_{STO,NO3},\ Y_{STO,NO2},\ \eta_{H,NO3},\ \eta_{H,NO2},\ K_{H,O2:inh},\ K_{H,NO3},\ K_{AOB,O2},\ K_{NOB,O2}]$ selected by the proposed NOAP could achieve a better fitting, with the same GO algorithm.

2.3 DISCUSSION

In this work, a novel NOAP approach for the efficient calibration of activated sludge models with limited available data has been proposed. The proposed NOAP integrates a new numerical global parameter sensitivity analysis method (DGSM) for factor fixing and a numerical pseudo-global parameter correlation analysis method for non-identifiable parameter detection to determine the optimal parameter subset for model calibration. The validity and applicability of the approach for efficient model calibration is confirmed by two different activated sludge systems (SBR and CSTR systems). The model calibration results suggested that the optimal sensitive parameter subsets of the SBR and CSTR system are different despite with the same extended ASM3 model to calibrate. Even with the same biomass collected from a municipal WWTP, two SBR reactors finally result in different optimal parameter subsets due to different operational conditions. The results indicate that the parameter subsets determined by NOAP can tail with the state variation of the system. This outstands from the experience-based procedures in calibrating dynamic systems as activated sludge systems whose parameters and structure can vary gradually, which would facilitate modeling automation a lot to support more optimization applications of WWTP.

The optimal parameter subset is different and specific for various systems, because of differences in environmental conditions, influent characteristics, operation modes and biomass population. As a black-box method, conventional experience-based model calibration procedures construct a mapping relation between data and model parameter subset, in which mapping routines are based on experts' empirical knowledge. However, uncertainty would be inevitable due to the arbitrary subset selection. For example, these risks may be distinctly enlarged when modeling the dynamic SBR scenario. An efficient calibration procedure is not only simple to fit the trend of historical data by manually selecting a parameter subset, but should be competent for optimal parameter subset determination, with the aids of efficient parameter estimation algorithms. The proposed NOAP in this study could be a promising alternative to fulfill the described demands. Since the global sensitivity analysis possesses the ability to evaluate uncertainty impact of the concerning factors on the model outputs scientifically. In fact, through factors fixing, parameter estimation is performed for top uncertainty introducers, which would reduce outputs' uncertainty in maximization. Consequently, through parameter estimation of subset determined by the global sensitivity analysis, the possibility of successful model calibration and prediction can be maximized. Furthermore, parameter subset optimized by numerical global parameter correlation analysis would enhance the success of calibration and validation. Simultaneously, the proposed NOAP method can quickly capture the shifts of system states through continuous updating of the known Factors' values and unknown Factors' boundary variations. Thus, the proposed NOAP is a promising and useful tool for the efficient calibration of ASMs and could potentially apply to other ordinary differential equations models. To overcome the time varying character, the global sensitivity analysis indices of each parameter are defined as equation (6).

Currently, the proposed NOAP procedure is a decision-helping tool, rather than an automatic protocol. In fact, it is expected to develop a fully automatic calibration procedure in future. Firstly, the automatic calibration procedure should recognize the optimal parameter subset for any models, and organize efficient parameter estimation automatically and robustly, through efficient numerical or symbolic algebra calculation approaches. Secondly, automatic methods are available for optimal experiments design and data collection, when uncertainty analysis of the parameter estimated is necessary in case of improving confidence intervals. Moreover, this calibration procedure can provide implementation procedure for automatic modification, selection and extension of model structure, when both model calibration and validation failed after data optimization. Most importantly, the calibrated model cannot only monitor and predict the overall process dynamics, but also facilitate the n~1 n~1 operators to achieve optimal control of a target system.

2.4 METHODS

2.4.1 PARAMETER ESTIMATION, PARAMETER SENSITIVITY AND CORRELATION ANALYSIS

The parameter estimation is a critical step of the model calibration process. Stochastic global optimization algorithms can find the global minimum of the objective function given by equation (1) [48,49], where $y_0 = [COD(0)$ $S_{NH4}(0) \, S_{NO2}(0) \, S_{NO3}(0)]^T$ as the example. Fitting goodness criterion function for each observable is given by equation (2). Characterizations and explanations of the symbols presented in this section can be found in the Table S1 (Supporting Information, SI).

$$E(\theta)= \sum_{k=1}^{N_i} [y^5(t_k)-y^0(t_k)]^T \omega^{-1}(t_k)[y^5(t_k)-y^0(t_k)]$$

Table 1: Parameter estimation results of each model calibrations after global optimization using Genetic Algorithm

Subset suggested by reference Value			Subset selected by NOAP First calibration of SBR scenario		Re-calibration of SBR for improvement		CSTR scenario	
Parameter	SBR scenario	C S T R scenario	Parameter	Value	Parameter	Value	Parameter	CSTR scenario
μ_{AOE}	0.78 d⁻¹	0.38 d⁻¹	$Y_{STO,O2}$	0.53 g COD/g COD	$Y_{STO,O2}$	0.53 g COD/g COD	$Y_{STO,NO3}$	0.50 g COD/g COD
μ_{NOE}	0.73 d⁻¹	0.34 d⁻¹	$Y_{STO,NO3}$	0.44 g COD/g COD	$Y_{STO,NO3}$	0.44 g COD/g COD	$Y_{STO,NO2}$	0.36 g COD/g COD
$\eta_{H,NO3}$	0.11	0.08	$Y_{STO,NO2}$	0.24 g COD/g COD	$Y_{STO,NO2}$	0.24 g COD/g COD	$\eta_{H,NO3}$	0.42
$\eta_{H,NO2}$	0.99	0.94	k_{STO}	4.37 d⁻¹	k_{STO}	4.37 d⁻¹	$\eta_{H,NO2}$	0.52
$\eta_{H,endNO3}$	0.86	0.05	μ_{AOB}	1.02 d⁻¹	μ_{AOB}	1.02 d⁻¹	$K_{H,O2,inh}$	0.73 g O₂ m⁻³
$\eta_{H,endNO2}$	0.95	0.98	μ_{NOB}	0.76 d⁻¹	μ_{NOB}	0.76 d⁻¹	$K_{H,NO3}$	5.69 g N m⁻³
$(K_{H,O2,inh})$	(18.44 g O₂ m⁻³)	-	$\eta_{H,NO2}$	0.78 d⁻¹	$K_{H,O2,inh}$	(2.85 g O₂ m⁻³)	$K_{AOB,O2}$	3.11 g O₂ m⁻³
-	-	-	-	-	$\eta_{H,NO2}$	0.78	$K_{NOB,J2}$	2.47 g O₂ m⁻³

where

$$\frac{dy}{dt}=f(y,x,u,\theta)$$

$$y(0)=y_0$$

$$A_i(B_i)=1-\frac{\sum_{k=1}^{N_i}w^{-1}(y_i^5(t_k)-y_i^0(t_k))^2}{\sum_{k=1}^{N_i}w^{-1}(\overline{y}_i^0(t_k)-y_i^0(t_k))^2}$$

GA possesses the advantages of easy implementation and mature codes to reuse compared to other resembled technologies [50–52]. In this study, the MATLAB R2010a (Global Optimization Toolbox) is referred as the numerical function implementation of GA (The Mathworks Inc. USA).

To realize factor fixing, a Derivative based Global Sensitivity Measures (DGSM) method is introduced to perform the global sensitivity analysis [53,54]. By comparisons among DGSM, Morris and Sobol's method, it indicates that: a) DGSM shows much higher convergence rate and more accurate than Morris method for non-monotonic functions; b) there is a link between DGSM and Sobol's global sensitivity indices, but the computational time required for numerical evaluation of DGSM measures is many orders of magnitude lower [55,56]. Essentially, the DGSM method is based on the local sensitivity measure, but perform an average of the local sensitivity measure throughout the parameter space by introducing Quasi Monte Carlo sampling methods. The relative time varying sensitivity matrix is described as the following equation (3).

$$\overline{s}_{ij}=\frac{\theta_j}{y_i}(\frac{\partial y_i}{\partial \theta_j})_{y=y(t,\hat{\theta}),\theta=\hat{\theta}}$$

Average $\overline{S}_{\parallel}$ over the parameter space using Quasi Monte Carlo sampling methods, a measure can be defined as the equation (4).

$$\overline{M}_{ij} = \int_{H^{\wedge}\theta} \overline{S}_{ij} d\theta$$

The numerical computation format can be expressed as the equation (5).

$$\overline{M}^*_{ij} = \frac{1}{N_\theta} \sum_{n=1}^{N_\theta} | \overline{S}_{ij} | d\theta$$

To overcome the time varying character, the global sensitivity analysis indices of each parameter are defined as equation (6).

$$\overline{M}^*_j = \frac{1}{N_y} \frac{1}{N_s} \sum_{i=1}^{N_y} \sum_{k=1}^{N_s} \overline{M}^*_{ij}(t_k)$$

About the global Correlation Analysis, a pseudo-global correlation matrix is introduced [54]. The local Fisher Information Matrix (FIM) is described as equation (7).

$$FIM = \sum_{k=1}^{N_s} [\frac{\partial y^s(t_k)}{\partial \theta}]^T w_i^{-1}(t_k)[\frac{\partial y^s(t_k)}{\partial \theta}]$$

The derivative covariance matrix is an approximation of the inverse of the FIM as equation (8).

$$C = FIM^{-1} = [\sum_{k=1}^{N_s} [\frac{\partial y^s(t_k)}{\partial \theta}]^T w_i^{-1}(t_k)[\frac{\partial y^s(t_k)}{\partial \theta}]]^{-1}$$

To introduce the pseudo-global covariance matrix, the local covariance matrix needs to be averaged throughout the parameter space like DGSM done with each objective function's value as the weight as equation (9).

$$\overline{C}_{ij} = \frac{1}{\sum_{n=1}^{N_v} E(\theta_n)} \sum_{n=1}^{N_v} C_{ij}(\theta_n) E(\theta_n)$$

According to the pseudo-global covariance matrix, the correlation matrix is defined as equation (10).

$$\left\{ \begin{array}{l} R_{ij} = \dfrac{\overline{C}_{ij}}{\sqrt{\overline{C}_{ii}\,\overline{C}_{jj}}}, i \neq j; \\ \\ R_{ij} = 1, i = j \end{array} \right.$$

Based on the equations (6) and (10), parameter sensitivity ranking order and correlation relationships would be produced systematically.

2.4.2 ACTIVATED SLUDGE MODEL AND EXPERIMENTAL DATA FOR NOAP TESTING

An extended ASM3 for two-step nitrification and denitrification [6] is used for verifying the proposed procedure. The model inherits the basic mechanism settings of ASM3, in the frame of "Hydrolysis – Storage – Growth - Respiration", nitrification and denitrification are extended to meet current need of description for main intermediate product – nitrite. The kinetic equations and stoichiometric matrix are presented in Table S2 and Table S3 (SI), respectively. Model structure and parameter settings are kept as the original for possibility of results comparison.

In addition, basic stoichiometric and kinetic parameters related information is presented in the Table S4 (SI), as well as parameter boundaries for GSA and parameter estimation. The validity and applicability of the approach is confirmed using experimental data obtained with two independent wastewater treatment systems, including SBR [45] and CSTR [46], respectively.

Experimental data of the SBR related scenario were collected from two reactors with a working volume of 14 L. Both reactors were seeded with the same inoculum from a full-scale municipal WWTP, but operated in different modes [45]. One was operated with the complete nitrification mode, while the other one was operated with the partial nitrification mode. The complete nitrification mode was operated in the aerobic-anoxic scheme with extensive aeration. Each cycle of the aerobic-anoxic scheme consisted of 3 min feeding, aeration, anoxic phase, 1 h settling, 6 min decanting, and 1 min idling. Aeration was still provided for another 0.5 h after the ammonium has been completely oxidized to nitrite, which would offer ideal environment for the nitrite-oxidizing bacteria to oxidize the nitrite successively to nitrate completely. The data from the complete nitrification reactor are used for the preliminary model calibration because the kinetic properties of the microorganisms in the system can be properly captured by these data series. In addition, the reactor with partial nitrification mode was also operated in the aerobic-anoxic scheme, but aeration duration was controlled through a real-time control system. The data from the partial nitrification reactor with obvious nitrite accumulation are applied to validate the preliminary model calibration results [45].

The lab-scale CSTR was set up to achieve partial nitritation. The reactor had an effective reaction volume of 4 L, followed by an clarifier with a working volume of 3.5 L. Sludge retention time (SRT) was kept at 12 days by wasting sludge from the secondary clarifier. Experimental data of the CSTR related scenario illustrated an obvious effect of DO on the nitrite accumulation in the CSTR system, which could be used to identify and estimate DO related switching function parameters. DO play an important role in biological nitrogen removal processes. Controlling DO at a proper level can not only reduce energy consumption, but also favor the partial nitrification for nitrogen removal via nitrite [57]. During the simulation in this case, experimental data from 113-day operation of the CSTR are divided into two groups. One group (0th to 75th day) is used for the preliminary calibration, while the other group (75th to 113th day) is used for the validation.

REFERENCES

1. Hauduc, H. et al. Critical review of activated sludge modeling: State of process knowledge, modeling concepts, and limitations. Biotechnol and Bioeng 110, 24–46 (2013).
2. Hauduc, H. et al. Activated sludge modelling in practice: an international survey. Water Sci Technol 60, 1943–1951 (2009).
3. Gernaey, K. V., van Loosdrecht, M. C. M., Henze, M., Lind, M. & Jorgensen, S. B. Activated sludge wastewater treatment plant modelling and simulation: state of the art. Environ Modell Softw 19, 763–783 (2004).
4. Rieger, L., Koch, G., Kuhni, M., Gujer, W. & Siegrist, H. The Eawag Bio-p Module for Activated Sludge Model NO. 3. Water Res 35, 3887–3903 (2001).
5. Iacopozzi, I., Innocenti, V., Marsili-Libelli, S. & Giusti, E. A modified Activated Sludge Model No.3 (ASM3) with two-step nitrification-denitrification. Environ Modell Softw 22, 847–861 (2007).
6. Kaelin, D., Manser, R., Rieger, L., Eugster, J., Rottermann, K. & Siegrist, H. Extension of ASM3 for two-step nitrification and denitrification and its calibration and validation with batch tests and pilot scale data. Water Res 43, 1680–1692 (2009).
7. Garcia-Usach, F., Ribes, J., Ferrer, J. & Seco, A. Calibration of denitrifying activity of polyphosphate accumulating organisms in an extended ASM2d model. Water Res 44, 5284–5297 (2010).
8. Ni, B. J., Ruscalleda, M., Pellicer-Nacher, C. & Smets, B. F. Modeling nitrous oxide production during biological nitrogen removal via nitrification and denitrification: Extensions to the general ASM models. Environ Sci Technol 45, 7768–7776 (2011).
9. Nopens, I. et al. Model-based optimisation of the biological performance of a side-stream MBR. Water Sci Technol 56, 135–143 (2007).
10. Houweling, D., Wunderlin, P., Dold, P., Bye, C., Joss, A. & Siegrist, H. N2O Emissions: Modeling the Effect of Process Configuration and Diurnal Loading Patterns. Water Environ Res. 83, 2131–2139 (2011).
11. Koch, G., Kuhni, M., Gujer, W. & Siegrist, H. Calibration and Validation of Activated Sludge Model no. 3 for Swiss Municipal Wastewater. Water Res 34, 3580–3590 (2000).
12. Penya-Roja, J. M., Seco, A., Ferrer, J. & Serralta, J. Calibration and validation of Activated Sludge Model No.2d for Spanish municipal wastewater. Environ Technol 23, 849–862 (2002).
13. Ludwig, T. et al. An advanced simulation model for membrane bioreactors: development, calibration and validation. Water Sci Technol 66, 1384–1391 (2012).
14. Wett, B. et al. Models for nitrification process design: one or two AOB populations? Water Sci Technol 64, 568–578 (2011).
15. Dochain, D. & Vanrolleghem, P. A. Dynamical Modelling and Estimation in Wastewater Treatment Processes. IWA Publishing (2001).
16. Petersen, B., Gernaey, K., Devisscher, M., Dochain, D. & Vanrolleghem, P. A. A simplified method to assess structurally identifiable parameters in Monod-based activated sludge models. Water Res 37, 2893–2904 (2003).

17. Zhang, T., Zhang, D. J., Li, Z. L. & Cai, Q. Evaluating the structural identifiability of the parameters of the EBPR sub-model in ASM2d by the differential algebra method. Water Res 44, 2815–2822 (2010).

18. Checchi, N., Giusti, E. & Marsili-Libelli, S. PEAS: A toolbox to assess the accuracy of estimated parameters in environmental models. Environ Modell Softw 22, 899–913 (2007).

19. Nopens, I., Hopkins, L. N. & Vanrolleghem, P. A. An overview of the posters presented at Watermatex 2000. III. Model selection and calibration/optimal experimental design. Water Sci Technol 43, 387–389 (2001).

20. Sharifi, S., Murthy, S., Takacs, I. & Massoudieh, A. Probabilistic parameter estimation of activated sludge processes using Markov Chain Monte Carlo. Water Res 50, 254–266 (2014).

21. Brun, R., Kuhni, M., Siegrist, H., Gujer, W. & Reichert, P. Practical identifiability of ASM2d parameters - systematic selection and tuning of parameter subsets. Water Res 36, 4113–4127 (2002).

22. Ruano, M. V., Ribes, J., De Pauw, D. J. W. & Sin, G. Parameter subset selection for the dynamic calibration of activated sludge models (ASMs): experience versus systems analysis. Water Sci Technol 56, 107–115 (2007).

23. Kim, Y. S., Kim, M. H. & Yoo, C. K. A new statistical framework for parameter subset selection and optimal parameter estimation in the activated sludge model. J Hazard Mater. 183, 441–447 (2010).

24. Mannina, G., Cosenza, A., Vanrolleghem, P. A. & Viviani, G. A practical protocol for calibration of nutrient removal wastewater treatment models. J Hydroinform 13, 575–595 (2011).

25. Sin, G. et al. Modelling nitrite in wastewater treatment systems: a discussion of different modelling concepts. Water Sci Technol 58, 1155–1171 (2008).

26. Sin, G., Vanhulle, S., Depauw, D., Vangriensven, A. & Vanrolleghem, P. A critical comparison of systematic calibration protocols for activated sludge models: A SWOT analysis. Water Res 39, 2459–2474 (2005).

27. Gillot, S., Ohtsuki, T., Rieger, L., Shaw, A., Takacs, I. & Winkler, S. Development of a unified protocol for good modeling practice in activated sludge modeling. Influents 4, 70–72 (2009).

28. Rieger, L. et al. Guidelines for Using Activated Sludge Models. 25–26 (IWA Publishing, London, 2013).

29. Machado, V. C., Tapia, G., Gabriel, D., Lafuente, J. & Baeza, J. A. Systematic identifiability study based on the Fisher Information Matrix for reducing the number of parameters calibration of an activated sludge model. Environ Modell Softw 24, 1274–1284 (2009).

30. Makinia, J. & Wells, S. A. A general model of the activated sludge reactor with dispersive flow - II. Model verification and application. Water Res 34, 3997–4006 (2000).

31. Makinia, J., Rosenwinkel, K. H. & Spering, V. Long-term simulation of the activated sludge process at the Hanover-Gummerwald pilot WWTP. Water Res 39, 1489–1502 (2005).

32. Gujer, W. Systems Analysis for Water Technology. [8–10] (Springer, Verlag Berlin Heidelberg, 2008).

33. Saltelli, A. et al. Global Sensitivity Analysis. The Primer. [10–39] (John Wiley & Sons, West Sussex, 2008).
34. Neumann, M. B. Comparison of sensitivity analysis methods for pollutant degradation modelling: A case study from drinking water treatment. Sci Total Environ 433, 530–537 (2012).
35. Cosenza, A., Mannina, G., Vanrolleghem, P. A. & Neumann, M. B. Global sensitivity analysis in wastewater applications: A comprehensive comparison of different methods. Environ Modell Softw 49, 40–52 (2013).
36. Morris, M. D. Factorial sampling plans for preliminary computational experiments. Technometrics 33, 161–174 (1991).
37. Cukier, R. I., Fortuin, C. M., Shuler, K. E., Petschek, A. G. & Schaibly, J. H. Study of the sensitivity of coupled reaction systems to uncertainties in rate coefficients. I. Theory. J Chem Phys 59, 3873–3878 (1973).
38. Schaibly, J. H. & Shuler, K. E. Study of the sensitivity of coupled reaction systems to uncertainties in rate coefficients. II. Applications. J Chem Phys 59, 3879–3888 (1973).
39. Cukier, R. I., Schaibly, J. H. & Shuler, K. E. Study of the sensitivity of coupled reaction systems to uncertainties in rate coefficients.3. Analysis of the approximations. J Chem Phys 63, 1140–1149 (1975).
40. Saltelli, A., Tarantola, S. & Chan, K. P. S. A quantitative model-independent method for global sensitivity analysis of model output. Technometrics 41, 39–56 (1999).
41. Sobol, I. M. Sensitivity estimates for nonlinear mathematical models. Math Modelling Comput Experiment 1, 407–414 (1993).
42. Chis, O, Banga, J. R. &, Balsa-Canto, E. GenSSI: a software toolbox for structural identifiability analysis of biological models. Bioinformatics 27, 2610–2611 (2011).
43. Vajda, S., Godfrey, K. & Rabitz, H. Similarity transformation approach to identifiability analysis of nonlinear compartmental models. Math Biosci 93, 217–248 (1989).
44. Chis, O. T., Banga, J. R. & Balsa-Canto, E. Structural Identifiability of Systems Biology Models: A Critical Comparison of Methods. PloS ONE 6, e27755 (2011).
45. Yang, Q., Liu, X. H., Peng, C. Y., Wang, S. Y., Sun, H. W. & Peng, Y. Z. N_2O production during nitrogen removal via nitrite from domestic wastewater- main sources and control method. Environ Sci Technol 43, 9400–9406 (2009).
46. Peng, Y. Z., Guo, J. H., Horn, H., Yang, X. & Wang, S. Y. Achieving nitrite accumulation in a continuous system treating low-strength domestic wastewater: switchover from batch start-up to continuous operation with process control. Appl Microbiol Biot 94, 517–526 (2012).
47. Corominas, L. et al. New framework for standardized notation in wastewater treatment modelling. Water Sci Technol 61, 841–857 (2010).
48. Moles, C. G., Mendes, P. & Banga, J. R. Parameter estimation in biochemical pathways: A comparison of global optimization methods. Genome Res 13, 2467–2474 (2003).
49. Chou, I. C. & Voit, E. O. Recent developments in parameter estimation and structure identification of biochemical and genomic systems. Math Biosci 219, 57–83 (2009).
50. Kim, S., Lee, H., Kim, J., Kim, C., Ko, J. & Woo, H. Genetic algorithms for the application of Activated Sludge Model No.1. Water Sci Technol 45, 405–411 (2002).
51. Fang, F., Ni, B. J. & Yu, H. Q. Estimating the kinetic parameters of activated sludge storage using weighted non-linear least-squares and accelerating genetic algorithm. Water Res 43, 2595–2604 (2009).

52. Keskitalo, J. & Leiviska, K. Application of evolutionary optimisers in data-based calibration of Activated Sludge Models. Expert Syst Appl 39, 6609–6617 (2012).
53. Kucherenko, S., Rodriguez-Fernandez, M., Pantelides, C. N. S. Monte carlo evaluation of derivative based global sensitivity measures. Reliab Eng Syst Safe 94, 1135–1148 (2009).
54. Rodriguez-Fernandez, M. & Banga, J. R. SensSB: A software toolbox for the development and sensitivity analysis of systems biology models. Bioinformatics 26, 1675–1676 (2010).
55. Kucherenko, S., Rodriguez-Fernandez, M., Pantelides, C. & Shah, N. Monte Carlo evaluation of derivative-based global sensitivity measures. Reliab Eng Syst Safe 94, 1135–1148 (2009).
56. Sobol, I. M. & Kucherenko, S. Derivative based global sensitivity measures and their link with global sensitivity indices. Math Comput Simulat 79, 3009–3017 (2009).
57. Guo, J. H., Peng, Y. Z., Wang, S. Y., Zheng, Y. A., Huang, H. J. & Wang, Z. W. Long-term effect of dissolved oxygen on partial nitrification performance and microbial community structure. Bioresour Technol 100, 2796–2802 (2009)

There are several supplemental files that are not available in this version of the article. To view this additional information, please use the citation on the first page of this chapter.

CHAPTER 3

Sorption and Release of Organics by Primary, Anaerobic, and Aerobic Activated Sludge Mixed with Raw Municipal Wastewater

OSKAR MODIN, SOROUSH SAHEB ALAM, FRANK PERSSON, AND BRITT-MARIE WILÉN

3.1 INTRODUCTION

Modern wastewater treatment is dominated by the activated sludge process, which was developed over 100 years ago [1]. Organic compounds are biologically oxidized in an aerated tank. Then, the sludge is typically separated from the treated effluent in sedimentation tanks, partly returned to the inlet of the aerated tank, and partly wasted as excess sludge (or waste activated sludge, WAS) [2]. Alternatively, membrane filtration can be used to separate the activated sludge from the treated water [3].

In many existing and emerging activated sludge process configurations, rapid sorption of organic compounds from the wastewater onto the sludge plays an important role for the removal. Existing process configurations that rely on sorption as a major removal mechanism include con-

Sorption and Release of Organics by Primary, Anaerobic, and Aerobic Activated Sludge Mixed with Raw Municipal Wastewater. © _Modin O, Saheb Alam S, Persson F, Wilén B-M. PLoS ONE **10,3** (2015). doi: 10.1371/journal.pone.0119371. Licensed under a Creative Commons Attribution 4.0 International License, http://creativecommons.org/licenses/by/4.0/._

tact-stabilization [4] and adsorption-biooxidation (AB) [5,6]. In the contact-stabilization process, the influent wastewater is mixed with activated sludge in a contact tank having a short hydraulic retention time (HRT) of e.g. 15 min. Organic compounds are assumed to rapidly sorb onto the sludge, which is separated from the treated water in a sedimentation tank. The settled sludge is then aerated in a stabilization tank to oxidize the sorbed organics, before being recycled back to the contact tank. The AB process is a two-sludge system consisting of a high-rate activated sludge process (the A-stage) operated with a short solids retention time (SRT) of 3–12 hours. In the A-stage, organics are removed mainly by sorption onto the sludge. The low-loaded B-stage is then used e.g. for nutrient removal and oxidation of the organic compounds remaining in the wastewater after the A-stage [7,8].

Directing WAS to the primary settlers to be mixed with the primary sludge in a wastewater treatment plant is another quite common practice to utilize the sorptive capacity of the WAS and improve the dewatering properties of secondary sludge [2,9,10]. However, scientific studies on the effect of activated sludge addition on organics removal in primary sedimentation are scarce. Yetis and Tarlan [9] found that addition of WAS improved sedimentation of suspended solids in raw wastewater under certain conditions. Tests were carried out with sludge cultivated at different solids retention times and generally concentrations above 1600 mg total suspended solids (TSS)/L gave the best results. However, no information was provided about dissolved substances. Ross and Crawford [11] carried out full-scale tests comparing primary settlers that either received or did not receive WAS. However, they did not see any significant differences in the organics content in the effluent from the settlers.

New activated sludge processes based on sorption as the major mechanism for organics removal are also emerging. As energy-efficiency and carbon footprint are becoming more and more important for wastewater treatment plants [12], high-rate activated sludge processes similar to the A-stage of an AB process are being investigated [13,14]. A high-rate activated sludge process with short solids retention time (SRT) is potentially more energy-efficient than a low-rate process with long SRT because less oxygen is needed per mass of organic material removed and more excess sludge is produced, which can be converted to biogas in an anaerobic digester. Thus, a high-rate process could both cut electricity consumption because of lower aeration requirements and increase energy output (in the form of produced biogas) for a wastewater treatment plant [15].

Sorption-based processes have also been investigated as an enhanced form of primary treatment. Huang and Li [16] recycled primary sludge and mixed it with the raw wastewater before primary sedimentation. Enhanced organics removal could only be obtained after the primary sludge had been aerated in a stabilization tank. A COD removal efficiency of 40% could be obtained which was 35% higher than with primary sedimentation alone. Zhao et al. [17] investigated a bioflocculation-adsorption, sedimentation and stabilization process for enhanced primary treatment. They obtained total COD removal efficiencies of 70–80%. Both of the processes referred to here are very similar to the contact-stabilization process.

To design and predict the performance of sorption-based activated sludge processes for enhanced primary treatment or for more energy-efficient wastewater treatment, studies investigating the sorption capacity of sludge are needed. The goal of this study is to quantify the sorption of particulate and dissolved organic compounds onto different types of sludge available at wastewater treatment plants. We compare primary, anaerobic, and aerobic activated sludge as sorbents for wastewater organics. For aerobic activated sludge, we also investigate the effect of starvation on sorption capacity, determine sorption kinetics, and investigate some characteristics of the sorbed organics.

3.2 MATERIALS AND METHODS

3.2.1 COLLECTION OF SLUDGE AND WASTEWATER

Activated sludge, primary sludge, anaerobic digester sludge, raw municipal wastewater, and treated effluent samples were collected at the Rya wastewater treatment plant, which treats about 4.4 m^3/s of municipal wastewater from the city of Gothenburg, Sweden. Permission to take samples at the plant was granted by Gryaab. The activated sludge basins at the plant have a solids retention time of 3–5 days and consist of an anoxic pre-denitrification zone followed by an aerobic zone. The basins are aerated using a diffused aeration system. Activated sludge samples were collected near the outlet of the aerobic zone. Primary sludge refers to sludge collected from the primary settlers. Anaerobic sludge was collected from a mesophilic digester treating sludge generated at the plant. The raw municipal wastewater refers to the influent to the plant before it had passed any treatment steps. The treated effluent refers to the final effluent from the plant collected after disc filtration, which is the final treatment operation. Starved activated sludge, which was used in some experiments, was obtained by keeping 8-L activated sludge in an aerated tank stirred at 200 rpm using a 4-bladed propeller (10 cm diameter) for 1, 3 and 6 days without the addition of substrate. All experiments and analyses were carried out less than 8 hours from the time samples were collected at the plant.

3.2.2 SORPTION TESTS

In the sorption tests, 100 mL of sludge suspension was mixed with 600 mL of raw municipal wastewater in 1-L beakers containing paddles (width 5.5 cm, height 3 cm) stirring the mixtures at 200 rpm. The velocity gradient in the beakers was estimated to 4.3 s^{-1} (the calculation is described in S1 File). After 5 min, mixing was stopped and the sludge-wastewater mixtures were allowed to settle for 30 min. Then, approximately 100 mL of the supernatant was collected for analyses. The sludge suspensions were prepared by centrifugation of the desired sludge volume at 1300g, decan- tation of the supernatant, and resuspension in 100 mL of effluent water. In control tests, 100 mL of effluent without added sludge was added to the 600 mL of wastewater. Measurements of conductivity, total suspended solids (TSS) and volatile suspended solids (VSS) concentrations, sludge volume (SV) and sludge volume index (SVI), particulate organic carbon (TOCp) and dissolved organic carbon (TOCd) concentrations, and absorbance measurements at 650 nm (ABS650) and 254 nm (ABS254) were carried out to characterize the samples and measure removal of organic compounds. All experiments were carried out in room temperature (about 22°C).

3.2.3 STATISTICAL METHODOLOGY FOR THE SORPTION TESTS

On each sampling day, the removal efficiencies of particulate and dissolved organic carbon were compared under three conditions: (1) addition of treated effluent (control), (2) addition of low concentration of activated sludge (0.30–0.39 gTSS/L), and (3) addition of high concentration of activated sludge (1.04–1.27 gTSS/L). The concentrations of added sludge refer to the final concentrations in the wastewater and sludge suspension mixtures. Duplicate tests were carried out for each condition. Paired-sample t-tests were conducted on the removal efficiencies for the different conditions. The null hypothesis was that there would be no difference in the removal efficiency with and without sludge addition. The null hypothesis was rejected if the two-tailed p-value was less than 0.05.

3.2.4 KINETIC TESTS

To determine the rate of sorption, kinetic tests were carried out using similar conditions to the sorption tests described above. After 1, 5, 10, 15, 30, 60, and 120 min of mixing, 50 mL of solution was withdrawn from each beaker and allowed to settle for 30 min in a separate vial. A sample of the supernatant was analyzed for ABS650 as a proxy for TSS concentration. A centrifuged sample (5 min at 4000g) was analyzed for ABS254 as a proxy for TOCd concentration [18]. Repeated measurements of ABS650, ABS254, TSS, and TODd of several wastewater and sludge samples from seven different sampling occasions showed strong correlations between ABS650 and TSS concentration ($R^2 = 0.99$), and ABS254 and TOCd concentration ($R^2 = 0.94$) (S2 File). On one of the test days, high-performance size exclusion chromatography (HPSEC) was carried out to investigate the fate of organic molecules of the size range 100–20 000 Da during the kinetic sorption tests.

The sorption was modelled using 1st order kinetics, which has previously been observed to describe sorption of particulate chemical oxygen demand (COD) onto activated sludge [19] (Equation 1).

$$dC/dt = -kxXx(C-a)$$

where dC/dt is the rate of change of organic sorbate concentration, C, with time, t (mg/ L·min), k is the first order rate constant (L/mgVSS·min), a is the residual concentration (mg/ L), and X is the activated sludge concentration (mgVSS/L).

The rate constant, k, and residual concentration, a, were found using Excel Solver by fitting the experimental data to Equation 2.

$$C(t) = a + (C_0-a)xexp(-kxXxt)$$

where C(t) is the organic sorbate concentration at time t, and C_0 is the initial concentration.

3.2.5 AZIDE INHIBITION TESTS

To quantify the effect of microbial respiration on the measured sorption rates of dissolved organics, experiments with sodium azide-inhibited sludge was carried out. The experiments were performed in a similar way as the kinetic tests with sampling after 1, 5, 10, 15, 30, and 60 min of mixing. Before the experiments, the sludge was exposed to approximately 0.2 gNaN$_3$/gTSS for three hours [20], which is a load known to inhibit the respiratory activity of activated sludge [21]. Sodium azide can cause deflocculation of sludge; therefore, controls with addition of azide-inhibited sludge to beakers containing treated effluent were run in parallel to quantify the release of organic substances by the sludge. Controls with live activated sludge were also run in parallel to allow direct comparison between azide-inhibited and live sludge.

3.2.6 ANALYTICAL METHODS

Conductivity was measured using a probe (WTW TetraCon325). TSS and VSS concentrations were analyzed according to Standard Methods [22] using glass-fiber filter papers (Munktell, grade MGA). SV was analyzed by allowing the sludge to settle in a 1 L graduated cylinder and measuring the volume occupied by the settled sludge after 30 min. SVI was obtained by dividing the SV by the TSS concentration. TOCp and TOCd concentrations were analyzed by a total organic carbon analyzer (TOC-V, Shimadzu). The TOC was divided into TOCp and TOCd fractions based on filtration through a 0.45 μm membrane-filter. Cake layer formation on the membrane during filtration could potentially result in retention of smaller particles than 0.45 μm. To minimize this effect, samples were centrifuged for 2 min at 4000 g before being filtered. Absorbance at 650 nm (ABS650) and at 254 nm (ABS254) was measured with a spectro-photometer (UV-1800, Shimadzu). Specific UV absorbance (SUVA) was calculated as the ratio between ABS254 and TOCd concentration. SUVA is an indicator of the aromatic content of the dissolved organic carbon [23]. HPSEC was carried out using a Shimadzu HPLC system equipped with UV and refractive index detectors, an Agilent Bio SEC-5 column (length 300 mm, diameter 7.8 mm, pore size 100Å), and a mobile phase made up of 100 mM NaCl, 8.3 mM KH2PO4, and 11.7 mM K_2HPO_4 being pumped at 0.5 mL/min. The molecular sizes of the eluted molecules were calibrated against retention time using polyethylene glycol standards (PEG-10 calibration kit, Agilent). The organic compounds in the sorption test samples were analyzed using the UV detector at wavelengths of 215, 230, 254, and 280 nm. Low molecular weight carboxylic acids (C1–C5) were analyzed using the same HPLC system equipped with an Aminex HPX-87H column (BioRad) and operated with a mobile phase of 5 mM H_2SO_4 pumped at 0.5 mL/min. The acids were detected at 210 nm.

3.3 RESULTS

3.3.1 GENERAL

Tests were carried out with wastewater and sludge collected at the treatment plant on eight different days. In the influent wastewater, the conductivity ranged from 791 to 1022 μS/cm and the TSS concentration ranged from 0.109 to 0.926 g/L. The activated sludge had a TSS concentration of 2.08–2.73 g/L, a VSS/TSS ratio of 70–73%, and a SVI of 70–96 mL/ gTSS. The primary sludge and anaerobic digester sludge had TSS concentrations of approximately 47 g/L and 30 g/L, respectively. The treated effluent from the plant contained 11–13 mg/L of TOCd and negligible concentrations of TSS.

3.3.2 NET RELEASE OF ORGANICS BY PRIMARY AND ANAEROBIC SLUDGE

The original concentrations of particulate and dissolved organics and the final concentration after 5 min mixing and 30 min sedimentation in controls and tests with addition of primary- and anaerobic sludge are shown in Fig. 1. Tests with addition of primary sludge and anaerobic digester sludge showed a net release of organic compounds from the sludge into the wastewater. Compared to the controls, both anaerobic digester sludge and primary sludge released particles into the wastewater, which can be seen by the increase in ABS650 and TOCp values. The amount of released TOCp was related to the amount of added sludge. For anaerobic digester sludge it was 45.3–49.5 mgTOCp/gVSS, whereas primary sludge released 21.7–27.8 mgTOCp/ gVSS. For dissolved substances, anaerobic digester sludge released 4.3–6.3 mgTOCd/gVSS, whereas primary sludge did not have a net release of TOCd. However, the ABS254 and consequently the SUVA increased with both sludges. This indicates that aromatic organics, such as humic acids, were released by the sludge. In the primary sludge, this release was balanced by an uptake of other organics resulting in no net change in the TOCd concentration.

3.3.3 NET SORPTION OF ORGANICS BY AEROBIC ACTIVATED SLUDGE

The change in TOCp and TOCd concentrations, ABS254 and ABS650 for the sorption tests carried out with aerobic activated sludge on seven different sampling days are shown in Fig. 2. The influent wastewater characteristics varied between the different sampling days, especially in terms

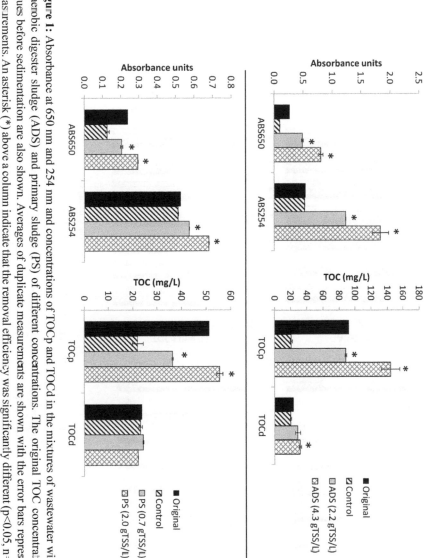

Figure 1: Absorbance at 650 nm and 254 nm and concentrations of TOCp and TOCd in the mixtures of wastewater with effluent (Control), anaerobic digester sludge (ADS) and primary sludge (PS) of different concentrations. The original TOC concentrations and absorbance values before sedimentation are also shown. Averages of duplicate measurements are shown with the error bars representing the individual measurements. An asterisk (*) above a column indicate that the removal efficiency was significantly different (p<0.05, n = 2) from the control.

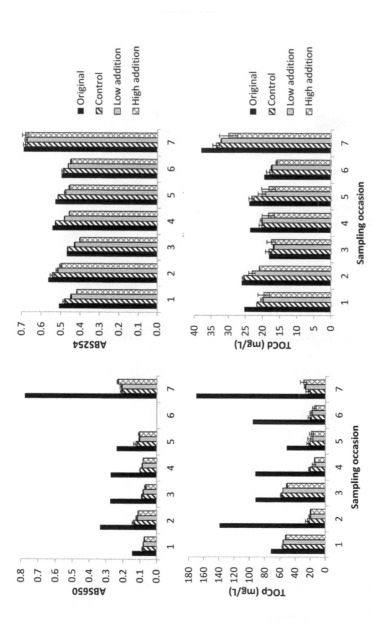

Figure 2: Absorbance at 650 nm and 254 nm and concentrations of TOCp and TOCd in the mixtures of wastewater with effluent or activated sludge suspension. Original refers to the concentrations before sedimentation. Control, low addition, and high addition refers to the concentration in the supernatant after 30 min sedimentation with zero (control), 0.30–0.39 gTSS/L (low addition), or 1.04–1.37 gTSS/L (high addition) activated sludge. Averages of duplicate measurements are shown with the error bars representing the individual measurements. ABS650 was not measured on sampling day 6.

of particulate content (see the original values of TOCp and ABS650 in Fig. 2). Addition of activated sludge generally appeared to have a small but positive impact on the net removal of both particulate (TOCp and ABS650) and dissolved (TOCd and ABS254) organic carbon. To verify that activated sludge addition did indeed have an effect on the removal, statistical analysis was carried out (Table 1). For the parameters ABS650, TOCp, ABS254, and TOCd addition of both low and high concentration of activated sludge always resulted in a statistically significant difference in removal efficiency compared to the controls suggesting that addition of activated sludge could improve removal of both particulate and dissolved substances. For all parameters except ABS650, there was also a significant difference in the removal between low (0.30–0.39 gTSS/L) and high (1.04–1.37 gTSS/L) activated sludge addition (Table 1). The average sorption of TOCp was 6.5±10.8 mgTOCp/gVSS while the sorption of TOCd was 5.0±4.7 mgTOCd/gVSS. Thus, the total organics sorption was 11.5±11.8 mgTOC/ gVSS. There was no difference in the SUVA between the controls and the samples with sludge addition suggesting that the aromatic content of the wastewater organics did not change because of mixing with activated sludge.

3.3.4 EFFECT OF ACTIVATED SLUDGE STARVATION

Starvation of activated sludge may free up adsorption sites, which would allow greater removal of organic compounds from wastewater [24]. Therefore, activated sludge was starved for 1, 3, and 6 days and its sorption ability was compared to fresh sludge collected from the wastewater treatment plant (Fig. 3). After 1 day of starvation, the starved sludge performed approximately equal to the fresh sludge, both resulting in an improvement of the removal of particulate and dissolved substances compared to the control. Only for ABS650 and addition of low concentration of sludge did the starved sludge show significantly worse removal efficiency than the fresh activated sludge. After 3 days starvation, high addition of sludge resulted in significantly lower removal efficiency of ABS650, TOCp, and UV254. After 6 days starvation, addition of both low and high concentrations of starved sludge resulted in significantly worse removal efficiencies than fresh activated sludge for all parameters except TOCd. Under no circumstances did starved sludge perform better than fresh activated sludge. The reason may be that the fresh activated sludge was collected from the outlet of the aeration tanks at the wastewater treatment plant. Thus, organic substances taken up by the sludge in the treatment plant had already been metabolized and further starvation did not free up any additional adsorption sites. Instead, prolonged starvation (>1 day) led to deflocculation of the sludge and release of particles (ABS650 and TOCp) and humic substances (UV254), which meant that addition of starved sludge to the wastewater had a negative effect on the removal efficiencies of these parameters.

3.3.5 SORPTION KINETICS WITH AEROBIC ACTIVATED SLUDGE

The sorption kinetics was investigated with raw wastewater and aerobic activated sludge collected on two different days. The changes in concentrations of TSS and TOCd are shown in Fig. 4 while kinetic coefficients and R^2-values for the fits are shown in Table 2. Upon mixing wastewater and sludge, an instantaneous increase in TSS concentration and decrease in TOCd concentration can be observed. The increase in TSS can be explained by release of small particles by the sludge. The decrease in TOCd suggests an instantaneous sorption of TOCd from the wastewater onto the sludge. Following this instantaneous sorption or release, sorption of both TSS and TOCd obeys 1st order kinetics.

3.3.6 EFFECT OF AZIDE INHIBITION

The rate of biosorption of dissolved organics was compared with live and azide-inhibited activated sludge. Azide treatment resulted in some release of organic substances by the sludge, ranging from 2.5 to 3.0 mgTOCd/ gVSS in the two tests. This is shown by the higher starting concentrations of TOCd in the sorption tests with azide-inhibited sludge (Fig. 5A) compared to live activated sludge (Fig. 5B). However, both live and azide-inhibited sludge showed a near-instantaneous sorption event upon mix-

Table 1. Two-tailed p-values for paired-sample t-tests comparing the percentage removal of ABS650, ABS254, TOCp, and TOCd in sedimentation of wastewater without addition activated sludge (control) and with low addition (0.30–0.39 gTSS/L) or high addition (1.04–1.37 gTSS/L).

	ABS650	TOCp	UV254	TOCd
Control vs low addition	2.2×10^{-3}	1.6×10^{-2}	4.6×10^{-6}	1.1×10^{-3}
Control vs high addition	2.5×10^{-3}	1.3×10^{-3}	5.1×10^{-6}	2.3×10^{-4}
Low vs high addition	1.3×10^{-1}	1.2×10^{-2}	3.6×10^{-5}	3.5×10^{-5}

p-values below 5×10^{-2} are deemed to indicate significant difference in removal and are marked in bold ($n = 14$).

ing with raw wastewater. The extent of near-instantaneous sorption was 2.0–2.1 mgTOCd/gVSS for live activated sludge and 1.0–1.3 mgTOCd/gVSS for azide-inhibited sludge. The rate of removal was 24–36 mg-TOCd/gVSSd for live activated sludge and 30–48 mgTOCd/gVSSd for azide-inhibited sludge. First-order kinetics could be used to describe the data with kinetic coefficients of 0.014–0.020 L/gVSSd for both the live and azide-inhibited sludge (Table 2). However, using zero-order reaction rates provided equally good fits and the rates of removal shown above were calculated from the linear decrease in TOCd concentrations.

3.3.7 FATE OF SMALL MOLECULES DURING SORPTION TESTS

The fate of small molecules (<20 kDa) in the sorption tests was investigated using HPSEC. In this size range, most of the organic molecules were smaller than 2 kDa (Fig. 6). Detection at wavelengths of 215, 230, 254, and 280 nm was carried out. However, the results at 215 and 230 nm were very similar to each other, and so were the results at 254 and 280 nm. Therefore, only the results at 215 and 254 nm are shown in Fig. 6. Organic compounds with functional groups such as carboxyls, carboxylates, aldehydes, ketones, and esters have absorption maxima near 200 nm [25], and these groups are likely detected in the HPSEC spectra at 215

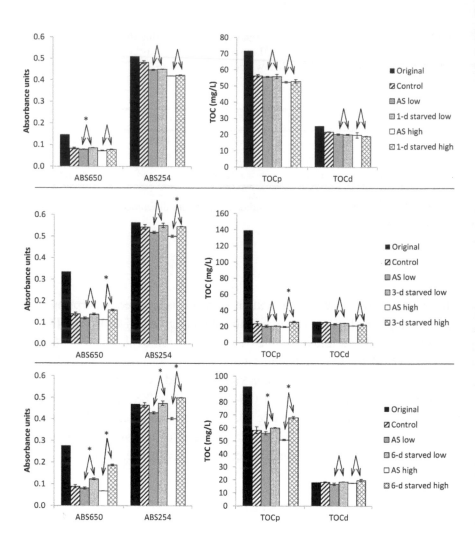

Figure 3: Absorbance at 650 nm and 254 nm and concentrations of TOCp and TOCd in the mixtures of wastewater with effluent (Control), 0.30–0.33 gTSS/L of activated sludge (AS low), 1.04–1.16 gTSS/L of activated sludge (AS high), or activated that had been starved for either 1, 3 or 6 days. The low concentration of the starved sludge was 0.28–0.32 gTSS/L and the high concentration was 0.97–1.13 gTSS/L. The original TOC concentration and absorbance values before sedimentation are also shown. Averages of duplicate measurements are shown with the error bars representing the individual measurements. Arrow pairs indicate that the removal efficiencies for two treatments were compared. An asterisk (*) above the arrow pair indicate statistically significant difference in removal ($p < 0.05$, $n = 2$).

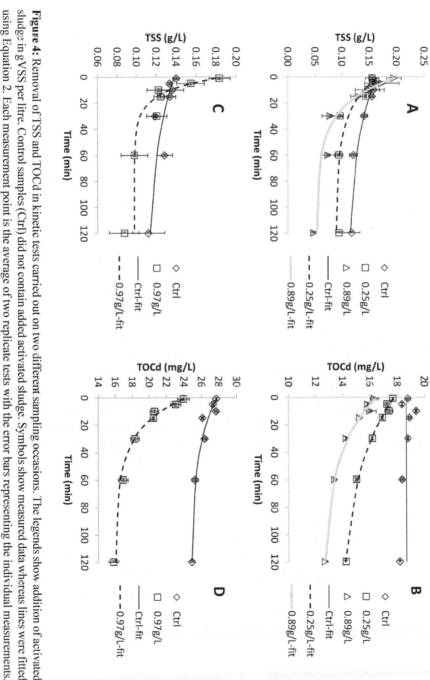

Figure 4: Removal of TSS and TOCd in kinetic tests carried out on two different sampling occasions. The legends show addition of activated sludge in gVSS per litre. Control samples (Ctrl) did not contain added activated sludge. Symbols show measured data whereas lines were fitted using Equation 2. Each measurement point is the average of two replicate tests with the error bars representing the individual measurements.

nm. At longer wavelengths, aromatic organic compounds have absorption peaks. Benzene and more complex aromatic molecules such as humic acids have absorption peaks near 254 nm and this wavelength is often used as an indicator of natural organic matter in water [26]. Proteins usually have absorption peaks near 280 nm because of the aromatic amino acids tryptophan and tyrosine [27]. Fig. 6 shows samples after 1 min of mixing, thus the effect of the near-instantaneous sorption event can be observed. At 215 nm, there is a clear difference between sorption tests and controls for molecules of 200 Da and less. This suggests that small molecules such as carboxylic acids may have been rapidly taken up by the sludge upon mixing. At 254 nm, there is no obvious difference between control and sorption test samples for molecules smaller than 20 kDa. As the TOCd values in the kinetic tests were obtained by correlation with ABS254 values, this suggests that the instantaneous sorption of TOCd observed in Fig. 4 is due to dissolved organic matter larger than 20 kDa. The HPSEC spectra obtained after mixing 15 and 120 min are shown in S3 File. For the short wavelengths (215 and 230 nm) the sorption tests samples are still distinctly different from the control around 200 Da at 15 min. However, at 120 min the absorbance at this size range has increased drastically both for the control and test samples, possibly because of fermentation leading to the production of carboxylic acids. At the longer wavelengths (254 and 280 nm), the absorbance of molecules smaller than 200 Da decreases with time in the sorption test samples whereas the control samples remain fairly constant. At 120 min, there is a clear difference between the controls and the test samples. This suggests that prolonged mixing with sludge leads to uptake or conversion of the aromatic fraction of organic molecules smaller than 200 Da.

Analysis with HPLC of low molecular weight organic acids (i.e. formic-, acetic, propionic-, lactic-, butyric-, crotonic-, and valeric acid) showed that neither of these were produced during the sorption tests. Thus, the increase in the UV absorbance (215 and 230 nm) of low molecular weight compounds after 120 min of mixing in the sorption tests was either caused by carboxylic acids larger than C5 or other small organic molecules containing functional groups absorbing at those wavelengths.

Table 2: Instantaneous sorption onto aerobic activated sludge, residual non-sorbable concentration (a), 1st order sorption constant (k), and coefficient of determination (R^2) for the kinetic tests.

Sample	Instant sorption	a	k	R^2
TSS sorption	(mgTSS/gVSS)	(mgTSS/L)	(L/gVSS.min)	
July 2013: 0.25 gVSS/L	-14	109	0.159	0.90
July 2013: 0.89 gVSS/L	-45	89	0.053	0.98
Oct. 2014: 0.97 gVSS/L	-51	53	0.094	0.91
TOCd sorption	(mgTOCd/gVSS)	(mgTOCd/L)	(L/gVSS.min)	
July 2013: 0.25 gVSS/L	8.7	18.7	0.065	0.99
July 2013: 0.89 gVSS/L	3.9	13.6	0.028	0.98
Oct. 2014: 0.97 gVSS/L	3.7	12.5	0.045	0.98
[a]Nov. 2014: 1.51 gVSS/L	2.0–2.1	12.7–13.7	0.014–0.020	0.95–0.98
[a]Nov. 2014: 1.51 gVSS/L	1.0–1.3	18.2–19.1	0.016–0.020	0.82–0.97

[a]**Test with** azide-inhibited activated sludge.

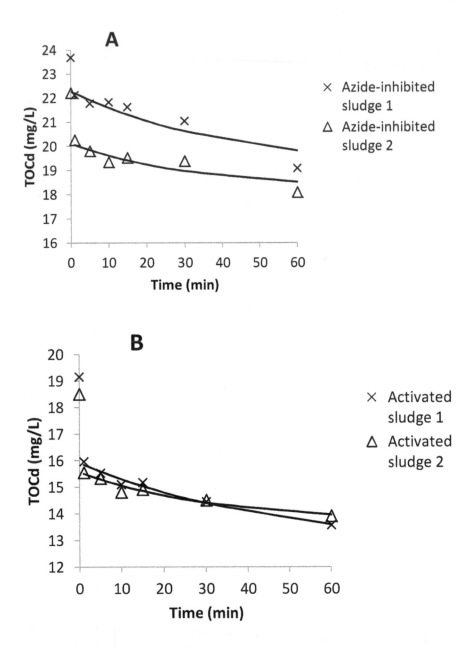

Figure 5: Concentration of TOCd in sorption tests with azide-inhibited (A) and live activated sludge (B). The solid lines show fits using 1st-order kinetics for the measurements points between 1 min and 60 min. Two repeated tests were carried out with each type of sludge.

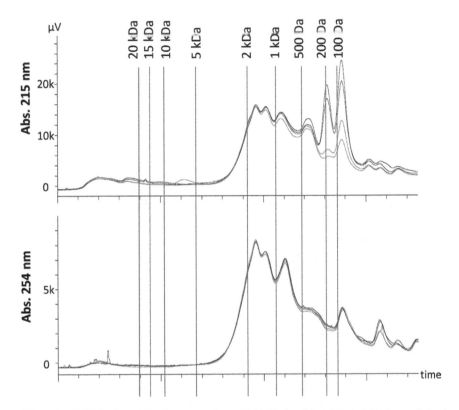

Figur 6: HPSEC of samples from sorption tests with aerobic activated sludge and 1 min mixing. Black lines show controls without addition of sludge. Red lines show samples with addition of 0.97 g/L VSS. The vertical lines show the retention time of polyethylene glycol standards of known molecular weight.

3.4 DISCUSSION

Three types of sludges were tested for sorption of organic compounds from raw municipal wastewater. Both primary- and anaerobic sludge had a net release of particles (>0.45 μm). Release of soluble substances (<0.45 μm) was lower; however, increases in the ABS254 values suggested that humic-like substances were released from the sludges. In this study, anaerobic sludge from a completely mixed mesophilic digester was used. It is possible that anaerobic granular sludge could have a net sorption of organics because of different surface characteristics and better settling properties than non-granular anaerobic sludge. Riffat and Dague [28] showed that anaerobic granules could adsorb milk from a synthetic solution. The sorption capacity ranged from less than 10 up to about 45 mgCOD/gTSS. This is similar to the 5-min sorption capacity of aerobic activated sludge observed in our study.

Recirculation of primary sludge to enhance primary sedimentation of wastewater has been suggested as an enhanced primary treatment technology [16]. However, in that study they observed that sorption of organics from the raw wastewater could only be obtained if the sludge was aerated in a separate tank before being contacted with the raw wastewater. Thus, the primary sludge evolved into an aerobic activated sludge. This is consistent with results in our study, which showed that addition of primary sludge did not remove organics from the wastewater, but addition of aerobic activated sludge did.

The 5-min sorption tests with aerobic activated sludge showed average sorption capacities of 6.5±10.8 mgTOCp/gVSS for particulates. Assuming a COD/TOC conversion factor of 2.67 (see S4 File), this is equivalent to 18±29 mgCOD/gVSS. This is similar to the 0–30 mgCOD/ gTSS sorption capacity observed by Rensink and Donker [29] for total wastewater COD in 10 min tests and the -40 to 100 mgCOD/gTSS sorption of non-settleable organics observed by Guellil et al. [30] in 60 min tests. The sorption capacity of 5.0±4.7 mgTOCd/gVSS for dissolved organics is also similar to the 0.9–16 mgTOCd/gTSS observed by Jorand et al. [31] in 15-min tests. Variation in sorption capacity between the different tests could depend on variations in both the activated sludge and the wastewater characteristics. The activated sludge appeared to have quite constant characteristics during the experimental period with SVI ranging from 70 to 96 mL/gTSS. The influent wastewater had more variable characteristics, which can be seen by the original values for ABS650, ABS254, TOCp, and TOCd in Figs. 1–3. However, no clear correlations could be found between any single of these parameters and sorption capacity. Instead, it is likely a combination of several different properties of the sludge and the wastewater that determine sorption capacity. Lim et al. [20] showed that operational parameters affecting sludge characteristics such as dissolved oxygen concentration, SRT, and pH can affect sorption capacity.

Sorption kinetics of wastewater organics onto sludge has previously been investigated in very few studies. Jimenez et al. [19] obtained a 1st order rate coefficient (k) of about 0.15 L/ gTSS·min for non-settlable particulates (>0.45 μm), which can be compared to the 0.05–0.16 L/gVSS·min observed in our study. For organics smaller than 0.45 μm the k-values were 0.030– 0.049 L/gTSS·min in Jimenez et al. [19] and 0.014–0.065 L/ gVSS·min in our study. The observations that both sorption capacity and kinetic coefficients correspond quite well with values obtained from other studies carried out in different countries suggest that the sorption capacity and kinetics of aerobic activated sludge is quite consistent between different treatment plants.

New high-rate activated sludge processes that rely on sorption as a major removal mechanism for organic matter are emerging. For example, Faust et al. [14] operated a high-rate MBR and estimated the degree of mineralization of organics from as low as 1% at 0.2 d to 11% at 1 d SRT. This means that energy can be saved by minimizing the need for aeration and energy can be recovered by maximizing the fraction of the organics that e.g. can be converted to biogas. To be able to model and optimize this and other types of sorption-based process, further knowledge about the mechanisms and rates of sorption is needed. Currently, sorption of organics onto activated sludge is often assumed to be an instantaneous process. For example, this is the assumption made (primarily for particulate and colloidal substrate) in the commonly used activated sludge models developed by the International Water Association [32]. Other studies have pointed out that sorption of organics is not instantaneous but can be modelled using kinetic equations [19], which is especially relevant for high-rate processes operated at short HRTs. In this study, we observed both a near-instantaneous sorption event and a slower sorption process obeying 1st order kinetics. The instantaneous sorption event occurred in less than 1 min of mixing, which was the time for our first sample. Tests with azide-inhibited sludge suggested that neither the instantaneous nor the slower sorption process was coupled to microbial respiration.

Further knowledge is needed about which fractions of the wastewater organics can be sorbed onto activated sludge. In all sorption tests, a significant residual organics concentration was observed in solution. HPSEC results showed that small molecular weight compounds (<200 Da) with strong UV absorbance at 215 and 230 nm were instantaneously taken up by the sludge upon mixing. However, the HPSEC spectra at 254 nm suggested that most of the organic compounds instantaneously sorbed were likely larger than 20 kDa. In a study by Dulekgurgen et al. [33], 43% of the dissolved COD in a municipal wastewater was associated with the size range 10 kDa-0.45 μm. Organics in this size range could include e.g. proteins, polysaccharides and humic acids [34]. After the instantaneous sorption and up to a mixing time of 120 min, the content of aromatic organic compounds smaller than about 200 Da was gradually decreased either by uptake, sorption, or conversion by the activated sludge microorganisms.

3.5 CONCLUSIONS

The sorption and release of organic compounds by primary-, anaerobic-, and aerobic activated sludge when mixed with raw municipal wastewater was investigated. Only aerobic activated sludge showed net sorption of organics from the wastewater. Primary- and anaerobic sludge had both net releases of mainly particulate organics.

For aerobic activated sludge, the 5-min sorption capacity was 6.5±10.8 mgTOCp/gVSS for particulate organics and 5.0±4.7 mgTOCd/gVSS for dissolved organics. This is similar to values obtained by other researcher using activated sludge from other treatment plants. Prolonged starvation under aerobic conditions of the activated sludge (>1 d) did not improve sorption capacity, instead it led to deflocculation.

When activated sludge was mixed with wastewater, a near-instantaneous sorption of dissolved organics (<0.45 μm) could be observed followed by a slower sorption obeying 1st order kinetics. For particles, there was an instantaneous release of particles when sludge and waste-water was mixed, followed by sorption according to 1st order kinetics. The kinetic rate coefficient was 0.05–0.16 L/gVSS·min for particles and 0.014–0.065 L/gVSS·min for dissolved organics.

HPSEC results suggested that organic compounds larger than 20 kDa but smaller than 0.45 μm as well as small molecules (<200 Da) with UV absorbance at 215–230 nm were almost instantaneously removed from the wastewater when mixed with activated sludge.

REFERENCES

1. Arden E, Lockett WT. Experiments on the oxidation of sewage without the aid of filters. J Soc Chem Ind. 1914; 33: 523–539.
2. Metcalf & Eddy Inc., Tchobanoglous G, Burton FL, Stensel HD. Wastewater Engineering, Treatment and Reuse 4th Ed. McGraw-Hill; 2004.
3. Yamamoto K, Hiasa M, Mahmood T, Matsuo T. Direct solid-liquid separation using hollow fiber membrane in an activated sludge aeration tank. Water Sci Technol. 1989; 21: 43–54.
4. Ullrich AH, Smith MW. The biosorption process of sewage and waste treatment. Sewage Ind Waste. 1951; 23: 1248–1253.
5. Versprille A, Zuurveen B, Stein T. The A-B process: A novel wastewater treatment system. Water Sci Technol. 1985; 17: 235–246.
6. Boehnke B, Diering B, Zuckut SW. Cost-effective wastewater treatment process for removal of organics and nutrient. Water Eng Manag. 1997; 144(5): 30–35.
7. Boehnke B, Diering B, Zuckut SW. Cost-effective wastewater treatment process for removal of organics and nutrients. Water Eng Manag. 1997; 144(7): 18–21.
8. Boehnke B, Schulze-Rettmer R, Zuckut SW. Cost-effective reduction of high-strength wastewater by adsorption-based activated sludge technology. Water Eng Manag. 1998; 145: 31–34.
9. Yetis U, Tarlan E. Improvement of primary settling performance with activated sludge. Environ Technol. 2002; 23: 363–372. PMID: 12088362
10. Imhoff K. Two-stage operation of activated sludge plants. Sewage Ind Waste. 1951; 27: 431–433. Ross RD, Crawford GV. The influent of waste activated sludge on primary clarifier operation. J Water Pollut Control Fed. 1985; 57: 1022–1026.
11. Gustavsson DJI, Tumlin S. Carbon footprint of Scandinavian wastewater treatment plants. Water Sci Technol. 2013; 68: 887–893. doi: 10.2166/wst.2013.318 PMID: 23985520
12. Diamantis V, Eftaxias A, Bundervoet B, Verstraete W. Performance of the biosorptive activated sludge (BAS) as pre-treatment to UF for decentralized wastewater reuse. Biores Technol. 2014; 156: 314–321. doi: 10.1016/j.biortech.2014.01.061 PMID: 24525216
13. Faust L, Temmink H, Zwijnenburg A, Kemperman AJ, Rijnaarts HH. High loaded MBRs for organic matter recovery from sewage: effect of solids retention time on bioflocculation and on the role of extracellular polymers. Water Res. 2014; 56: 258–266. doi: 10.1016/j.watres.2014.03.006 PMID: 24695067
14. Constantine T, Houweling D, Kraemer JT. "Doing the two-step"—reduced energy consumption sparks renewed interest in multistage biological treatment. Proceedings of the Water Environ Fed. 2012; 5771–5783.

15. Huang JC, Li L. Enhanced primary wastewater treatment by sludge recycling. J Environ Sci Health 2000; A35: 123–145.

16. Zhao W, Ting YP, Chen JP, Xing CH, Shi SQ. Advanced primary treatment of waste water using a bio-flocculation-adsorption sedimentation process. Acta Biotechnologica 2000; 1: 53–64.

17. Brandstetter A, Sletten RS, Mentler A, Wenzel WW. Estimating dissolved organic carbon in natural waters by UV absorbance (254 nm). Z Pflanzenernähr Bodenk 1996; 159: 605–607. PMID: 12287983

18. Jimenez JA, La Motta EJ, Parker DS. Kinetics of removal of particulate chemical oxygen demand in the activated sludge process. Water Environ Res. 2005; 77: 437–446. PMID: 16274077

19. Lim C-P, Zhang S, Zhou Y, Ng WJ. Enhanced carbon capture biosorption through process manipula- tion. Biochem Eng J. 2015; 93: 128–136.

20. Barbot E, Seyssiecq I, Roche N, Marrot B. Inhibition of activated sludge respiration by sodium azide addition: Effect on rheology and oxygen transfer. Chem Eng J. 2010; 163: 230–235.

21. Clescerl LS, Greenberg AE, Eaton AD. Standard Methods for the Examination of Water and Wastewater, 20th edition. APHA, AWWA, WEF; 1998.

22. Weishaar JL, Aiken GR, Bergamaschi BA, Fram MS, Fujii R, et al. Evaluation of specific ultaviolet absorption as an indicator of the chemical composition and re-activity of dissolved organic carbon. Environ Sci Technol. 2003; 37: 4702–4708. PMID: 14594381

23. Tan KN, Chua H. COD adsorption capacity of the activated sludge—Its determination and application in the activated sludge process. Environ Monit Assess. 1997; 44: 211–217.

24. McConnell JS, McConnell RM, Hossner LR. Ultraviolet spectra of acetic acid, glycine, and glyphosate. Proceedings Arkansas Academy of Science 1993; 47: 73–76.

25. Korshin GV, Li C-W, Benjamin MM. Monitoring the properties of natural organic matter through UV spectroscopy: a consistent theory. Water Res. 1997; 31: 1787–1795.

26. Schmid F-X. Biological macromolecules: UV-visible spectrophotometry. Encyclopedia of Life Sciences: 1–4; 2001.

27. Riffat R, Dague RR. Laboratory studies on the anaerobic biosorption process. Water Environ Res. 1995; 67: 1104–1110.

28. Rensink JH, Donker HJGW. The effect of contact tank operation on bulking sludge and biosorption processes. Water Sci Technol. 1991; 23: 857–866.

29. Guellil A, Thomas F, Block J-C, Bersillon J-L, Ginestet P. Transfer of organic matter between wastewater and activated sludge flocs. Water Res. 2001; 35: 143–150. PMID: 11257868

30. Jorand F, Block J-C, Palmgren R, Nielsen PH, Urbain V, et al. Biosorption of wastewater organics by activated sludge. Recent Progres en Genie des Procedes 1995; 44: 61–67.

31. Henze M, Gujer W, Mino T, van Loosdrech MCM. Activated sludge model ASM1, ASM2, ASM2d, and ASM3. IWA Publishing; 2000. doi: 10.1007/s00449-010-0446-2 PMID: 20607300

32. Dulekgurgen E, Dogruel S, Karahan O, Orhon D. Size distribution of wastewater COD fractions as an index for biodegradability. Water Res. 2006; 40: 273–282. PMID: 16376405
33. Shon H-K, Kim S-H, Erdei L, Vigneswaran S. Analytical methods of size distribution for organic matter in water and wastewater. Korean J Chem Eng. 2006; 23: 581–591.

There are several supplemental files that are not available in this version of the article. To view this additional information, please use the citation on the first page of this chapter.

CHAPTER 4

Removal Mechanisms and Kinetics of Trace Tetracycline by Two Types of Activated Sludge Treating Freshwater Sewage and Saline Sewage

BING LI AND TONG ZHANG

4.1 INTRODUCTION

In recent years, the occurrence and fate of antibiotics in the environment has drawn great attention of researchers all over the world (Kümmerer 2001; Xiao et al. 2008). Although the antibiotic residues in the environment are at the subinhibitory concentrations, they are still considered to be emerging pollutants because antibiotics may result in the development/ maintenance/transfer/spread of antibiotic-resistant bacteria and antibiotic-resistant genes in the long term (Kim et al. 2005; Knapp et al. 2008; Martínez 2008).

Tetracyclines, which ranked the second in production and usage among all antibiotic classes worldwide, are widely used as human and veterinary

Removal Mechanisms and Kinetics of Trace Tetracycline by Two Types of Activated Sludge Treating Freshwater Sewage and Saline Sewage. © *Li B and Zhang T.*.Environmental Science and Pollution Research International *20,5 (2013), (doi:10.1007/s11356-012-1213-5). Licensed under a Creative Commons Attribution 2.0 Generic License, http://creativecommons.org/licenses/by/2.0/.*

medicine as well as growth promoter (Gu and Karthikeyan 2005). However, tetracyclines are poorly metabolized or absorbed in the digestive tract and 50–80 % is excreted through feces and urine as unchanged form (Sarmah et al. 2006). For the human-use portion, wastewater treatment plants (WWTPs) are one of the dominant sources of tetracyclines which are released into the environment through effluent and biosolids as fertilizer because WWTPs cannot remove tetracyclines completely (Miao et al. 2004). For the animal-use portion, tetracyclines enter into the environment mainly through application of manure and waste lagoon water to fields as fertilizer (Boxall et al. 2004). Consequently, tetracyclines have been frequently detected in the environment, including effluent and sludge from WWTPs (Li et al. 2009; Miao et al. 2004; Spongberg and Witter 2008), surface water (Kim and Carlson 2007), sediment (Kim and Carlson 2007), and soils (Aga et al. 2005) around the world.

To date, although tetracyclines were found to be eliminated to some degree in the activated sludge process with the removal efficiencies of 11.6% (Spongberg and Witter 2008) to 85.4% (Batt et al. 2007), less attention was paid to their removal mechanisms (biodegradation, adsorption, volatilization, or hydrolysis) at environmentally relevant concentrations and the corresponding systematic studies were very limited. Understanding the removal of tetracycline by activated sludge is not only critical to the evaluation of tetracycline elimination in WWTPs but also to the mass load prediction and risk assessment/ management of tetracycline released to the soil environment since biosolids derived from WWTPs are widely applied to fields as fertilizer (Monteiro and Boxall 2009). Adsorption is an important process controlling the transport and fate of tetracyclines in the environment. Recent studies on adsorption of tetracyclines mainly focused on using isolated clays (Avisar et al. 2010; Chang et al. 2009), aluminum hydrous oxide (Gu and Karthikeyan 2005), soils (Sassman and Lee 2005; Wan et al. 2010), sediment (Xu and Li 2010), sand (Zhang et al. 2012), humic substances/clay–humic complexes (Pils and Laird 2007; Sun et al. 2010), and carbon nanotubes (Ji et al. 2010) as adsorbents. However, the removal behavior of tetracyclines obtained based on the above studies cannot be applied directly to activated sludge (AS) due to the vast difference between AS and those adsorbents as well as the solution chemistry conditions. In addition, the initial tetracycline concentrations for most previous studies were several orders of magnitude higher than their environmentally relevant concentrations, usually ranging from a few to hundreds milligrams per liter level. This might result in significant bias when predicting the removal behavior at environmentally relevant concentration levels. Moreover, the situation in Hong Kong is much more special and unique as some WWTPs (e.g., Shatin WWTP) treat saline sewage resulting from the practice of seawater toilet flushing. The constituents of the saline sewage are much more complicated than the freshwater sewage, and the different aqueous solution chemistry properties might lead to remarkably different removal behavior in AS process. To our knowledge, this is the first study to systematically examine the elimination of tetracycline, a principal member of tetracyclines, by two types of AS treating freshwater sewage and saline sewage, respectively, at environmentally relevant concentrations. Particular emphasis was placed on investigating (1) the removal mechanisms (biodegradation, adsorption, volatilization, and hydrolysis) for tetracycline, (2) adsorption kinetics and isotherms of tetracycline on AS, (3) the effect of ion species and ion concentrations on adsorption, and (4) the impact of pH and calculation of species-specific adsorption distribution coefficients.

4.2 MATERIALS AND METHODS

4.2.1 CHEMICALS AND STANDARDS

The standard of tetracycline (purity > 98%) was purchased from Sigma-Aldrich. LC-MS grade acetonitrile was purchased from Fisher Scientific UK Limited. Ultrapure water was prepared using Easypure® UV/UF compact reagent-grade water system (Barnstead, Boston, USA). The following chemicals were all purer than analytical grade: formic acid (99%) from Fluka, sodium hydroxide (>97%) from BDH, VWR International Ltd., hydrochloric acid (37%) from E. Merck, rhodamine B (~9%) and hexadecane (≥99%) from Sigma, individual standard solutions (1 g L^{-1}) of sodium, calcium, magnesium, chloride, and sulfate ions from Alltech Associates, Inc. (USA). Cellulose nitrate membrane (0.2 μm) was purchased from MFS® (Japan).

4.2.2 REMOVAL OF TETRACYCLINE IN ACTIVATED SLUDGE PROCESS

The saline sewage, freshwater sewage, and AS were collected from the aeration tanks (aerobic stage) of two local wastewater plants, i.e., Shatin and Stanley WWTPs, in August and September 2010, respectively. Table S1 (see Electronic supplementary material) shows the details of the two WWTPs. The major removal mechanisms for antibiotics in activated sludge process are considered to be biodegradation, adsorption, volatilization (due to aeration), and hydrolysis (Kim and Aga 2007; Pérez et al. 2005). In order to distinguish the primary removal mechanisms for tetracycline, batch test utilizing seven 2-L glass beakers with 1 L mixed liquor was run simultaneously at 25 ± 1 °C for 24 h. The detailed information on the experiment design was summarized in Table S2 and Section S 2.2 (see Electronic supplementary material).

4.2.3 CHARACTERIZATION OF SEWAGE

The sewages were first filtered using a 0.45-μm cellulose nitrate membrane at the sampling site, and then 50 mL filtrate was kept in an ice box and transported to the laboratory for the following analyses. The dissolved organic carbon (DOC) was measured using a total organic carbon analyzer (TOC-V$_{CPH,}$ Shimadzu, Japan). The cations (Na^+, Mg^{2+}, and Ca^{2+}) and anions (Cl^- and SO_4^{2-}) were analyzed by an ion chromatograph (COD-6A, Shimadzu, Japan), and the salinity was detected by a conductivity meter (Model 135A, ORION, USA).

4.2.4 CHARACTERIZATION OF AS

The characterization tests of two types of AS, i.e., zeta potential, particle size distribution, specific surface area, and relative hydrophobicity (RH), were conducted in duplicate within 10 h after sample collection.

4.2.4.1 ZETA POTENTIAL

The original sludge samples were first mixed thoroughly by a vortex mixer for 5 min for homogenization. Then, the supernatant was sampled after a 10-min settling for zeta potential measurement using DelsaTM Nano Series Zeta Potential Analyzers (DelsaTM Nano C, Beckman, USA) (Chang et al. 2001).

4.2.4.2 PARTICLE SIZE DISTRIBUTION

The size distribution of AS was determined by a laser diffraction particle size analyzer (LS™ 13 320, Beckman, USA). Both the mean and median values were utilized to characterize sludge size.

4.2.4.3 SPECIFIC SURFACE AREA

The specific surface area of AS was determined using the rhodamine B adsorption method which assumes that the adsorption isotherm fits the Langmuir model (Smith and Coackley 1983; Sørensen and Wakeman 1996). In this study, the adsorption isotherm test was conducted at 25 °C with the initial rhodamine B concentration of 0.5 to 75 mg L^{-1}. The AS concentration was 0.5 g SS L^{-1}, and the equilibrium time was 48 h. Rhodamine B was measured using a spectrophotometer (HACH, DR/2400) at the wavelength of 553 nm. Thus, the specific surface area can be calculated according to Eq. (1) (Smith and Coackley 1983):

$$S=q_{max}xNxA$$

where S is the specific surface area of the AS (in square meters per gram), q_{max} is the monolayer saturation adsorption capacity (in moles per gram) obtained from the Langmuir model fitting, N is Avogadro's number

$(6.023 \times 10^{23}$ molecules mol^{-1}), and A is the area occupied by a single rhodamine B molecule $(1.95 \times 10^{-18}$ m^2 molecule^{-1}) (Laurent et al. 2009).

4.2.4.4 RELATIVE HYDROPHOBICITY

The RH was evaluated following the protocol of Wilén et al. (2003). Thirty milliliters of AS mixed liquor was agitated uniformly with 15 mL hexadecane for 10 min in a separatory funnel. After a 30-min settling, the two phases were completely separated and the aqueous phase was transferred into another beaker. The RH was calculated using Eq. (2):

$$RH(\%) = (1 - \frac{MLSS_e}{MLSS_i}) x 100$$

where $MLSS_i$ and $MLSS_e$ are the AS concentrations in the aqueous phase before and after emulsification.

4.2.5 BATCH ADSORPTION STUDIES

All batch adsorption experiments were conducted in 100-mL conical flasks with rubber stoppers which were shaken on an orbital shaking incubator at 125 rpm in the dark to avoid possible photolysis of tetracycline. Stock solutions (100 mgL^{-1}) were freshly prepared daily by dissolving 10 mg tetracycline in 100 mL deionized water. About 20 L mixed liquor sampled from aeration tank was first settled by gravity for 30 min to concentrate the sludge and then the supernatant was filtered by a 0.45-μm cellulose nitrate membrane. The obtained "filtrate" (15 L) was kept at 4 °C before being used in the following batch adsorption experiments. Sodium azide was added into the concentrated sludge (~5 L) to get a final concentration of 1% (w/v) to inhibit the microbial activity by incubation for 48 h at room temperature (Batt et al. 2007). Then, 0.05 g sludge (MLSS) after the above treatment was collected by centrifugation at 4,000 rpm for 5 min and resuspended in 50 mL filtrate to achieve a final concentration of 2.5 gL^{-1}. Unless otherwise specified, the pH of the above suspension was adjusted to 7.00±0.05 with HCl or NaOH solutions and the initial tetracycline concentration was 100 µgL^{-1}. The temperature was controlled at 25±0.5 °C, and the equilibrium time of 24 h was chosen based on the adsorption kinetics study. Duplicate experiments were performed in parallel.

4.2.5.1 ADSORPTION KINETICS

Four groups of experiments were conducted in the kinetic study: (I) Stanley AS in the freshwater sewage filtrate (FSF), (II) Shatin AS in the saline sewage filtrate (SSF), (III) Stanley AS in SSF, and (IV) Shatin AS in FSF. Groups (I) and (II) represented the actual adsorption conditions in the AS process of Stanley and Shatin WWTPs, respectively. Groups (III) and (IV) were used to distinguish the key factors (AS properties and/or the sewage matrix) which lead to the different adsorption behavior in these two WWTPs. A series of sealed conical flasks containing the above 50 mL AS slurry with initial TC concentrations of 100 µgL^{-1} were shaken continuously on an orbital shaking incubator (125 rpm) at 25±0.5 °C in the dark. Blank samples without AS were kept in the same condition as the control. The conical flasks were taken out at different time intervals (2 min, 5 min, 10 min, 15 min, 30 min, 1 h, 2 h, 5 h, 10 h, 15 h, and 24 h) for subsequent tetracycline detection.

4.2.5.2 EFFECT OF ION SPECIES/CONCENTRATION

According to major ion species and concentrations detected in saline wastewater, NaCl, CaCl$_2$, MgCl$_2$, and Na$_2$SO$_4$ were all spiked into FSF to simulate the SSF, called as synthetic saline sewage (SSS). In addition, to distinguish the effects of different ions on tetracycline adsorption, the above salts were individually spiked into FSF at three concentration levels and the middle level was at their actual concentrations in SSF. All the experiments in this section used Shatin AS as the adsorbent. Blank samples without Shatin AS were used as the controls. The adsorption distribution coefficient K_d, which was used to examine the effect of the ion species and concentration on adsorption, is defined as follows:

$$K_d = \frac{q_e}{C_e} = \frac{(C_0 - C_e) \cdot V/M}{C_e}$$

where q_e (in micrograms per gram) is the amount of tetracycline adsorbed per gram sludge at equilibrium and C_e (in micrograms per liter) is the equilibrium aqueous tetracycline concentration.

4.2.5.3 EFFECTS OF PH

The suspension pHs were adjusted between 4.50 and 9.00 (0.50-unit increments) by adding small volumes of hydrochloric acid or sodium hydroxide solutions. Final pH values were measured at the equilibrium time of 24 h. Blank samples without AS were used as the controls.

4.2.5.4 ADSORPTION ISOTHERMS

Adsorption isotherms were obtained to assess tetracycline distributions between AS and aqueous phases as a function of the tetracycline concentration and temperature (Figueroa et al. 2004). The experiments were conducted at nine initial concentrations ranging from 1 to 100 $\mu g L^{-1}$ under three temperatures, i.e., 10, 25, and 35 °C, respectively. Both the concentration and temperature were selected at the environmentally relevant levels. Blank samples without AS were used as the controls for each concentration.

4.2.6 ULTRAPERFORMANCE LIQUID CHROMATOGRAPHY– TANDEM MASS SPECTROMETRY ANALYSIS

All the samples were filtered via a 0.2-µm cellulose nitrate membrane which showed no adsorption of tetracycline (Li et al. 2009), kept in dark at 4 °C, and analyzed directly via Acquity™ ultraperformance liquid chromatography–tandem mass spectrometry (UPLC-MS/MS, Waters) within 24 h. UPLC-MS/MS was operated in the positive electrospray ion- ization multiple reaction monitoring (MRM) mode. The tetracycline concentrations were quantified using external calibration method, and the standards were prepared using the corresponding sewage filtrate to correct the matrix effect. The limit of quantification of tetracycline was 0.05 µgL^{-1}. Sample pretreatment and mobile phase gradient were summarized in the Electronic supplementary material while other detailed information on MRM parameters, column, flow rate, and formic acid concentration were reported in the previous study (Li et al. 2009).

4.3 RESULTS AND DISCUSSION

4.3.1 REMOVAL OF TETRACYCLINE IN ACTIVATED SLUDGE PROCESS

As shown in Fig. S1 (see Electronic supplementary material), tetracycline was found to be stable and no hydrolysis occurred during the testing period, and the elimination due to volatilization can be ignored based on the data of treatments III and IV. Thus, only biodegradation and adsorption might account for the removal of tetracycline. However, the strong similarity between treatment I profile and treatment II profile suggests that adsorption is the primary mechanism for tetracycline removal in both fresh-water and saline sewage activated sludge systems while biodegradation can be completely ignored (Fig. S1). In addition, it should be noted that different adsorption rate and adsorption capacity were found in the two systems, and these will be discussed in detail in the following sections.

4.3.2 BATCH ADSORPTION STUDIES AND CHARACTERIZATION OF AS AND SEWAGE

4.3.2.1 ADSORPTION KINETICS

In general, three kinetic models, i.e., pseudo-first-order kinetics, pseudo-second-order kinetics, and Elovich model, were utilized to fit the adsorption data (Chang et al. 2009; Xu et al. 2009).

Pseudo-first-order kinetics can be expressed as the Lagergren's rate equation:

$$log(q_e - q_t) = log(q_e) - \frac{k_1}{2.303}t$$

Pseudo-second-order kinetics can be written as the following equation:

$$q_t = \frac{k_2 q_e^2 t}{1 + k_2 q_e t}$$

Equation (6) is the rearranged linear form:

$$\frac{t}{q_t} = \frac{1}{k_2 q_e^2} + \frac{1}{q_e}t$$

Elovich model is expressed as

$$q_t = a \ln t + b$$

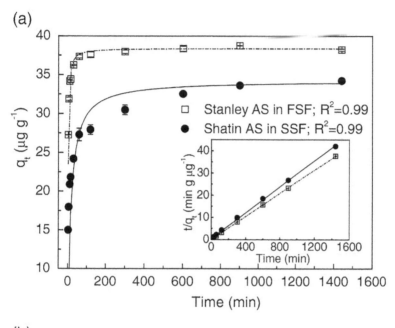

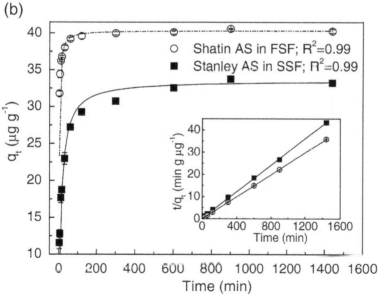

Figure 1: Adsorption kinetics (pseudo-second-order model) of tetracycline on AS. The inset is the linear plot of the pseudo-second-order model fit. Error bars mean SE. AS, activated sludge; FSF, freshwater sewage filtrate; SSF, saline sewage filtrate

Table 1: Pseudo-second-order kinetic model parameters for tetracycline adsorption to activated sludge

Group	Adsorption System	$C_0(\mu gL^{-1})$	$q_e(\mu gg^{-1})$	$q^*_e(\mu gg^{-1})$	$k_2(gmin^{-1}\mu g^{-1})$	R^2
I	Stanley AS in FSF	99.2	38.4	38.8	2.04×10^{-2}	0.99
II	Shatin AS in SSF	100.0	34.3	34.2	1.94×10^{-3}	0.99
III	Stanley AS in SSF	98.7	33.6	33.7	2.35×10^{-3}	0.99
IV	Shatin AS in FSF	104.3	40.3	40.6	1.70×10^{-2}	0.99

q_e calculated values based on pseudo-second-order kinetics model fitting, q^*_e experimental values where k_1 is the pseudo-first-order rate constant, k_2 is the pseudo-second-order rate constant, and a and b are the Elovich model constants.

Compared with the pseudo-first-order and Elovich models, the adsorption of tetracycline under all experimental conditions fitted the pseudo-second-order kinetic model best with $R^2 \geq 0.99$ (Table 1 and Table S3). This is in agreement with many other studies on the adsorption of tetracycline by various adsorbents, such as rectorite (Chang et al. 2009), calcined magnesium–aluminum hydrotalcites (Xu et al. 2009), and iron oxides–coated quartz (Tanis et al. 2008). As mentioned in "Adsorption kinetics," tetracycline adsorption by Stanley AS in FSF (group I) and by Shatin AS in SSF (group II) reflected the actual adsorption behavior in the AS processes of Stanley and Shatin WWTPs. As shown in Fig. 1a, after a very rapid adsorption during the first 0.5 and 2 h, complete equilibriums were reached in 2 and 15 h for AS processes of Stanley and Shatin WWTPs, respectively. Judging from the pseudo-second-order rate constants (k_2) of 2.04×10^{-2} and 1.94×10^{-3} gmin^{-1} μg^{-1} as well as the initial rate $k_2 q^2_e$ of 30.1 and 2.28 μgmin^{-1}g^{-1}, it could be concluded that the adsorption of tetracycline in Stanley WWTPs was much faster than that in Shatin WWTPs. Additionally, greater q_e suggested greater adsorption capacity and higher removal efficiency of tetracycline when the same initial concentration (100 μgL^{-1}) was applied in these two AS systems.

To distinguish the key factors (AS properties and/or sewage matrix) affecting the above different adsorption behavior, another two sets of experiments with the different combinations of AS and sewage types, i.e., Stanley AS in SSF (group III) and Shatin AS in FSF (group IV), were conducted. As shown in Fig. 1 and Table 1, similar adsorption behaviors (k_2 and q_e) were observed for group I and group IV, which represented two types of AS in the same sewage matrix (FSF). For group II and group III, similar adsorption trends (k_2 and q_e) were also found in these two AS systems with SSF as the sewage matrix. However, comparing the groups using the same AS but different sewage matrixes, i.e., group I and group III or group II and group IV, the adsorption phenomenon varied greatly. Thus, these results suggested that the sewage matrix (FSF or SSF) rather than the AS played a determinant role in the adsorption of tetracycline on AS. To confirm the above conclusion, the characterization of activated sludge and sewage was further conducted, respectively.

4.3.2.2 CHARACTERISTICS OF AS

The characteristics of AS from Stanley and Shatin WWTPs, including zeta potential, particle size distribution, specific surface area, and RH were summarized in Table 2. AS has a negative surface charge, indicated by the negative value of zeta potential (Kara et al. 2008). The zeta potential of Stanley AS and Shatin AS was −22.5 and −12.5 mV, respectively. A significantly higher zeta potential was observed for Shatin AS, possibly due to the presence of abundant cations in the saline sewage (Table 2), especially the bivalent cations (Ca^{2+} and Mg^{2+}), which may effectively lessen the negative surface charge of AS (Pevere et al. 2007). Both the mean and median size of Stanley AS were slightly smaller than those of Shatin AS while the specific surface areas of these two types of AS were almost the same. The RH is a suitable parameter to characterize the average hydrophobicity of the heterogeneous AS (Jin et al. 2003). In this study, Stanley AS and Shatin AS had RH values of 67 and 82 %, respectively, indicating the presence of both hydrophobic and hydrophilic groups at the sludge surface, and Shatin AS was slightly more hydrophobic than Stanley AS. Similar hydrophobicity results for AS (50–85 %) were reported by Jin et al. (2003) and Laurent et al. (2009).

4.3.2.3 AQUEOUS SOLUTION CHEMISTRY CHARACTERISTICS OF SEWAGE

The concentration of the cations (Na^+, Ca^{2+}, and Mg^{2+}), anions (Cl^- and SO_4^{2-}), DOC, and salinity of freshwater and saline sewage were also summarized in Table 2. Except for DOC, there was a vast difference for concentrations of cations, anions, and salinity between these two types of sewage. For Na^+, Mg^{2+}, Cl^-, and salinity, the concentrations in the saline sewage were 106–160 times higher than those in the freshwater sewage, while for Ca^{2+} and SO_4^{2-}, the concentrations in the saline sewage were about 10 and 27 times higher than those in the freshwater sewage. The comparison of the ion concentration as well as salinity between the saline sewage and seawater indicated that seawater accounted for a fraction of about 30% in the saline sewage.

Table 2: Characteristics of activated sludge and sewage from aeration tank

Property	Activated sludge	
	Stanley WWTP	Shatin WWTP
Zeta potential (mV)	-22.5±1.4	-12.5±0.7
Particle size distribution (μm)	Mean 106±0	Mean 137±0
	Median 80.9±0.1	Median 116±0
Specific surface area (m^2gSS^{-1})	28.2±5.3	28.6±4.5
RH%	6.7±1	82±0
Concentration (mgL^{-1})	Sewage from aeration tank	Saline sewage (Satin WWTP)
	Freshwater sewage (Stanley WWTP)	
Na^+	24.0±0.3	3,497±23
Ca^{2+}	11.5±0.2	116±0
Mg^{2+}	2.6±0.1	301±12
Cl^-	37.5±0.2	5,971±10
SO_4^{2-}	24.9±5.6	679±14
DOC	4.2±0.1	7.1±0.1
Salinity (ppt)	0.1±0.0	10.6±0.0

4.2.3.4 EFFECT OF ION SPECIES/CONCENTRATION

As mentioned above, the sewage matrix (FSF or SSF) rather than the AS played a determinant role in the different adsorption behavior of tetracycline on AS in freshwater and saline sewage systems, respectively. Adsorption distribution coefficient K_d was used to further examine the effect of the ion species and ion concentration on tetracycline adsorption (Sun et al. 2010; Ter Laak et al. 2006). As shown in Fig. 2, the K_d in FSF was 3.3 times as high as that in SSF while K_d in SSS was very similar to that in SSF. This indicated that the decrease of tetracycline adsorption in SSF might mainly result from the ions existed in SSF. This result is also in

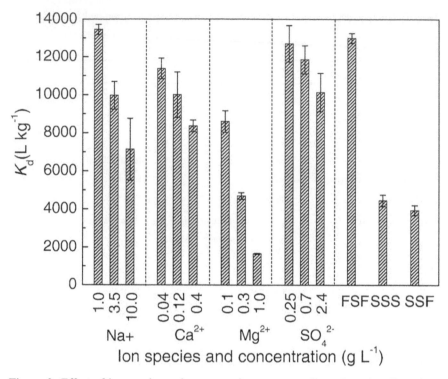

Figure 2: Effect of ion species and concentration on tetracycline adsorption. Error bars mean SE. FSF, freshwater sewage filtrate; SSS synthetic saline sewage; SSF, saline sewage filtrate

agreement with the previous study which reported that adsorption of tetracycline by marine sediment decreased with an increase of salinity (Xu and Li 2010).

However, salinity is a comprehensive parameter, and the previous studies did not specify the effect of each kind of ion on tetracycline adsorption. To distinguish the effects of different ions, NaCl, $CaCl_2$, $MgCl_2$, and Na_2SO_4 were respectively spiked into FSF at three concentration levels and their concentrations were calculated based on Na^+, Ca^{2+}, Mg^{2+}, and SO_4^{2-} as shown in Fig. 2. Judging from the K_d values at the middle concentration levels which were equal to their actual concentrations in SSF, it could be concluded that the decrease of adsorption in SSF compared to that in FSF was mainly due to the presence of Mg^{2+}. The impact of Cl^- was negligible because the corresponding Cl concentration (1.5 gL of 1.0 gL^{-1} Na$^+$ was much higher than that (0.89 gl^{-1}) of 0.3 gL^{-1}Mg^{2+} while the K$_d$ (similar to the K$_d$ in FSF) of the former was much greater than that of the latter. At the actual concentration level, the effect of Na$^+$ on decreasing tetracycline adsorption was comparable with that of Ca^{2+} but much smaller than that of Mg^{2+} although the mole concentration ratios of C$_{Na+}$/C$_{Ca2+}$ and C$_{Na+}$/C$_{Mg2+}$ were greater than 50 and 12, respectively. This indicated that the effect of monovalent Na$^+$ on adsorption was much less important than that of divalent Ca^{2+} and Mg^{2+}, and nonspecific electrostatic interactions should not be the predominant adsorption mechanism of tetracycline on AS (Tanis et al. 2008).

Tetracycline may form strong complexes with Ca^{2+} and Mg^{2+}, and the complexation of tetracycline with Ca^{2+} and Mg^{2+} in the aqueous phase might cause the decreased adsorption (Figueroa and Mackay 2005; Jin et al. 2007; Tanis et al. 2008). Additionally, the adsorption competition between positively charged quaternary ammonium functional group of tetracycline and divalent cations (Ca^{2+} and Mg^{2+}) for the cation exchange sites of adsorbent surface (e.g., carboxyl groups on extracellular polymeric substance of AS) will also decrease tetracycline adsorption (Liu et al. 2010; Pils and Laird 2007; Sun et al. 2009; Sun et al. 2010; Tanis et al. 2008). Comparing the K$_d$ values corresponding to Ca^{2+} of 0.4 gL^{-1} (0.01 M), Mg^{2+} of 0.1 gL^{-1} (0.004 M), and 0.3 gL^{-1} (0.0125 M), it can be found that the ability of Ca^{2+} to decrease tetracycline adsorption was weaker than that of Mg^{2+} given the same mole concentration. The K$_d$ values at three different concentrations demonstrated that the adsorption decreased with the increase of ion concentration and the trend was the most significant for Mg^{2+}. The occurrence of SO$_4^{2-}$ in SSF imposed a negligible impact on tetracycline adsorption.

4.2.3.5 EFFECT OF PH

As shown in Fig. 3, K$_d$ of tetracycline decreased with increasing pH over the tested pH range (4.5–8.4) and a gradual decrease trend was found between pH of 6.5 and 8.0 in both AS systems. Similar trend was also reported by other studies which investigated the pH effect on tetracycline

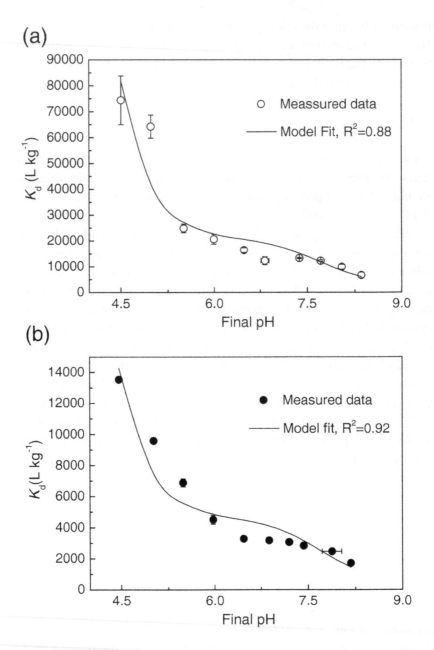

Figure 3: Effect of pH on K_d for tetracycline adsorption to AS: a Stanley AS in FSF and b Shatin AS in SSF. Error bars mean SE. AS, activated sludge; FSF, freshwater sewage filtrate; SSF, saline sewage filtrate

adsorption to montmorillonite and soils (Figueroa et al. 2004; Sassman and Lee 2005). For Shatin AS system, the K_d decreases more than 7.5 times over the whole experimental pH range, from 13,530±70 to 1,710±30 Lkg^{-1}. For the Stanley AS system, the K_d decreased more significantly, from 74,340±9,390 to 6,670±1,100 Lkg^{-1}, by approximately 11 times over a similar pH range. As illustrated in Fig. S4 (see Electronic supplementary material), tetracycline possesses multiple ionizable functional groups, i.e., tricarbonyl amide (C-1/C-2/C-3), phenolic diketone (C-10/C-11/ C-12), and dimethylamine (C-4) groups which correspond to three acid dissociation constants (pK$_a$ 0 3.3, 7.7, and 9.7), respectively. Thus, tetracycline exists as a cationic (+00), zwitterionic (+−0), and anionic (+−− or 0−−) species under acidic, moderately acidic to neutral, and alkaline conditions (Fig. S5, see Electronic supplementary material). It was reported that the ionization behavior is expected to significantly affect tetracycline sorption and each ionic species had its own adsorption magnitude (Figueroa et al. 2004; Gu and Karthikeyan 2005; Wang et al. 2008). To study the adsorption of different tetracycline species, the following empirical model, which expressed the overall adsorption distribution coefficient K_d as the sum of the species-specific adsorption distribution coefficients weighted with the corresponding fraction of individual species, was employed in this study.

$$K_d = K_d^{+00} \times f^{+00} + K_d^{+-0} \times f^{+-0} + K_d^{+--} \times f^{+--}$$

where K_d^{+00}; K_d^{+-0}, and K_d^{+--} are the species-specific adsorption distribution coefficients and f^{+00}, f^{+-0}, and f^{+--} are the fractions for cationic, zwitterionic, and anionic species, respectively. The anionic species (0−−) was neglected in the model due to its minor fraction in the tested pH range.

A nonlinear regression line was fitted to the K_d data at different pH values with Eq. (8) using Microsoft Office Excel 2003 software (Solver function). Well fits were obtained with the K_d^{+00}; K_d^{+-0}, and K_d^{+--} of 1.06× 10^6, 2.10×10^4, and 3.02×10^3 Lkg^{-1} in the Stanley AS system (R^2=00.88) while 1.54×10^5, 4.62×10^3, and 4.11× 10^2 Lkg^{-1} in the Shatin AS system (R^2=00.92), respectively. This suggested that the adsorption affinity of different tetracycline species with AS surface followed the order of cationic >> zwitterionic species>anionic species. The above model, Eq. (8), was

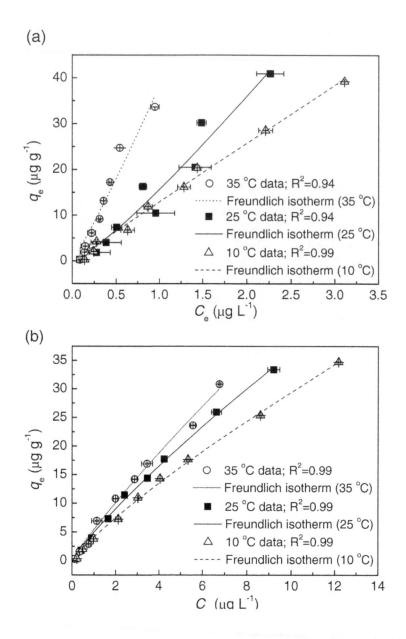

Figure 4: Adsorption isotherms of tetracycline on AS at 10, 25, and 35 °C: a Stanley AS in FSF (initial concentration 1–100 µgL−1), b Shatin AS in SSF (initial concentration 1–100 µgL−1), and c Shatin AS in SSF (initial concentration 100–5,000 µgL−1). Error bars mean SE. AS, activated sludge; FSF, freshwater sewage filtrate; SSF, saline sewage filtrate

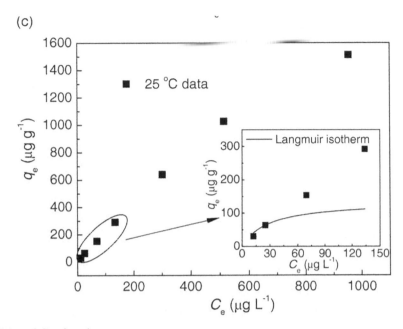

Figure 4 Continued

also used to fit tetracycline adsorption on soils (Sassman and Lee 2005) and montmorillonite (Figueroa et al. 2004). They found similar results indicating that cationic species had much higher affinity with adsorbents than zwitterionic species. The adsorption of anionic species was neglectable as K_d^{+--} was 0 (Sassman and Lee 2005). However, it was also reported that the adsorption of anionic tetracycline on montmorillonite was significant and K_d^{+--} was about half of K_d^{+-0} (Wang et al. 2008). By calculating the species-specific adsorption distribution coefficients weighted with the corresponding fraction, it was found that the contribution of zwitterionic tetracycline to the overall adsorption was always greater than 90% in the practical pH range of aeration tank (6.0–7.0). In the lower pH range (6.0–4.5), although the fraction of cationic species was as low as 0.2–6.4%, its contribution to the total adsorption ranged from 6.5 to 75% in both Shatin and Stanley AS systems. In the higher pH range (7.7–8.4), the fraction of anionic species was about 50 to 80% and its contribution to the total adsorption ranged 12–40% in these two AS systems.

4.2.3.6 ADSORPTION ISOTHERMS

Adsorption isotherms were utilized to assess tetracycline distributions between AS and aqueous phases as a function of tetracycline concentration. Freundlich and Langmuir adsorption isotherms were applied to fit the experimental data at three temperatures (10, 25, and 35 °C):

Freundlich model: $\quad q_e = K_f C_e^n$

Langmuir model: $\quad q_e = \dfrac{q_{max} b C_e}{1 + b C_e}$

where K_f (in microgram^{1-n} liter^n per gram) is the Freundlich affinity coefficient and gives an estimate of the adsorptive capacity, n (unitless) is the Freundlich linearity index, q_{max} (in micrograms per gram) is the maximum adsorption capacity, and b (in liters per microgram) is the Langmuir equilibrium coefficient.

The model fitting parameters were summarized in Table S4 (see Electronic supplementary material). Judging from the correlation coefficients R^2, it seems that the fitting performance of the Langmuir model was as good as that of the Freundlich model. However, the Freundlich rather than the Langmuir model was preferred as the applicable adsorption isotherm to fit the experimental data in this study due to the following reasons: Firstly, significantly inconsistent q_{max} and b values of the Langmuir model were obtained for Stanley AS system at different temperatures. For example, the q_{max} at 25 °C was 2 orders of magnitude greater than that at 10 °C, and this is unreasonable because temperature will not alter the maximum adsorption capacity of AS so much. Secondly, the predicted value of q_{max} based on the Langmuir model was not consistent with the actual situation. Taking Shatin AS system for example, the q_{max} of 135 µgg−1 at 25 °C suggested that the saturate adsorption amount of tetracycline on AS was 135 µgg^{-1} and it should not increase further even when the initial tetracycline concentration increased. However, as shown in Fig. 4c, the adsorption did not reach saturation at 135 μgg^{-1} and q_e raised from 30.4 to 1510 μgg^{-1} with the initial tetracycline concentration increasing from 100 to 5,000 μgL^{-1}. The possible reason might be that the basic assumption of the Langmuir adsorption model did not stand for the tested AS, that is, the adsorption of tetracycline on AS was multilayer instead of monolayer. Therefore, the Freundlich isotherm was applied to describe the adsorption of tetracycline on AS. In previous studies, the adsorption of tetracycline on soils (Wan et al. 2010), marine sediment (Xu and Li 2010), carbon nanotubes (Ji et al. 2010), and montmorillonite (Wang et al. 2008) was also found to follow the Freundlich isotherm better than the Langmuir isotherms.

Freundlich adsorption isotherm fittings at three temperatures for Stanley AS and Shatin AS were presented in Fig. 4a, b, respectively. Visually, higher temperature could enhance the adsorption process and the similar result was reported before (Tanis et al. 2008). In addition, K_f can be also regarded as a relative indicator of adsorption capacity, i.e., greater K_f indicated higher adsorption capacity of the adsorbent (Xu and Li 2010). When temperature increased from 10 to 35 °C, K_f raised from 12.9 to 38.2 (in micrograms^{1-n} liter^n per gram) and from 4.15 to 5.48 (in micrograms^{1-n} liter^n per gram) in these two adsorption systems, respectively. At the same temperature, the K_f of Stanley AS was approximately three times (10 and 25 °C) and seven times (35 °C) greater than the corresponding K_f of Shatin AS.

4.3 CONCLUSIONS

The removal mechanisms and kinetics of trace tetracycline, which were affected by aqueous solution chemistry properties of sewage and activated sludge surface characteristics, were systematically investigated in this study. The major conclusions were as follows:

- Adsorption is the primary removal mechanism for tetracycline in both freshwater and saline sewage activated sludge systems while biodegradation, volatilization, and hydrolysis can be completely ignored. This will enhance the transport of tetracycline into soil environment and increase the risk of development/maintenance/transfer/spread of tetracycline-resistant bacteria and tetracycline-resistant genes in the long term.

- Adsorption of tetracycline on AS fitted pseudo-second-order kinetics model well with $R^2 \geq 0.99$. Faster adsorption rate (k_2 2.04×10^{-2} g min^{-1} µg^{-1}) and greater adsorption capacity (q_e 38.8 µgg^{-1}) were found for freshwater AS than saline AS (k_2 1.94×10^{-3} gmin^{-1} µg^{-1}, q_e 34.3 µgg^{-1}).

- The Mg^{2+} in the saline sewage played a predominant role in decreasing tetracycline adsorption on saline AS.

- Adsorption of tetracycline decreased significantly with increasing pH over the tested pH range (4.5–8.4) and K_d was reduced from $13,530 \pm 70$ to $1,710 \pm 30$ Lkg^{-1} in saline AS system and from $74,340 \pm 9,390$ to $6,670 \pm 1,100$ L kg^{-1} in freshwater AS system. The adsorption affinity of different tetracycline species with AS surface followed the order of cationic species >> zwitterionic species > anionic species. Contribution of zwitterionic tetracycline to the overall adsorption was always greater than 90% over the actual pH range (6.0–7.0) in aeration tank.

- Adsorption of tetracycline in a wide range of temperature (10 to 35°C) followed the Freundlich adsorption isotherm well with R^2 ranging from 0.94 to 0.99.

REFERENCES

1.   Aga DS, O'Connor S, Ensley S, Payero JO, Snow D, Tarkalson D (2005) Determination of the persistence of tetracycline antibiotics and their degradates in manure-amended soil using enzyme-linked immunosorbent assay and liquid chromatography-mass spectrometry. J Agric Food Chem 53:7165–7171

2.   Avisar D, Primor O, Gozlan I, Mamane H (2010) Sorption of sulfonamides and tetracyclines to montmorillonite clay. Water Air Soil Pollut 209:439–450

3.   Batt AL, Kim S, Aga DS (2007) Comparison of the occurrence of antibiotics in four full-scale wastewater treatment plants with varying designs and operations. Chemosphere 68:428–435

4.   Boxall ABA, Fogg LA, Blackwell PA, Blackwell P, Kay P, Pemberton EJ, Croxford A (2004) Veterinary medicines in the environment. Rev Environ Contam Toxicol 180:1–91

5.   Chang GR, Liu JC, Lee DJ (2001) Co-conditioning and dewatering of chemical sludge and waste activated sludge. Water Res 35:786–794

6.   Chang PH, Jean JS, Jiang WT, Li ZH (2009) Mechanism of tetracycline sorption on rectorite. Colloids Surf A 339:94–99

7. Figueroa RA, Mackay AA (2005) Sorption of oxytetracycline to iron oxides and iron oxide-rich soils. Environ Sci Technol 39:6664–6671

8. Figueroa RA, Leonard A, Mackay AA (2004) Modeling tetracycline antibiotic sorption to clays. Environ Sci Technol 38:476–483

9. Gu C, Karthikeyan KG (2005) Interaction of tetracycline with aluminum and iron hydrous oxides. Environ Sci Technol 39:2660–2667

10. Ji LL, Chen W, Bi J, Zheng SR, Xu ZY, Zhu DQ, Alvarez PJ (2010) Adsorption of tetracycline on single-walled and multi-walled carbon nanotubes as affected by aqueous solution chemistry. Environ Toxicol Chem 29:2713–2719

11. Jin B, Wilén BM, Lant P (2003) A comprehensive insight into floc characteristics and their impact on compressibility and settleability of activated sludge. Chem Eng J 95:221–234

12. Jin LH, Amaya-Mazo X, Apel ME, Sankisa SS, Johnson E, Zbyszynska MA, Han A (2007) Ca^{2+} and Mg^{2+} bind tetracycline with distinct stoichiometries and linked deprotonation. Biophys Chem 128:185–196

13. Kara F, Gurakan GC, Sanin FD (2008) Monovalent cations and their influence on activated sludge floc chemistry, structure, and physical characteristics. Biotechnol Bioeng 100:231–239

14. Kim S, Aga DS (2007) Potential ecological and human health impacts of antibiotics and antibiotic-resistant bacteria from wastewater treatment plants. J Toxicol Environ Health B 10:559–573

15. Kim SC, Carlson K (2007) Quantification of human and veterinary antibiotics in water and sediment using SPE/LC/MS/MS. Anal Bioanal Chem 387:1301–1315

16. Kim S, Eichhorn P, Jensen JN, Weber AS, Aga DS (2005) Removal of antibiotics in wastewater: effect of hydraulic and solid retention times on the fate of tetracycline in the activated sludge process. Environ Sci Technol 39:5816–5823

17. Knapp CW, Engemann CA, Hanson ML, Keen PL, Hall KJ, Graham DW (2008) Indirect evidence of transposon-mediated selection of antibiotic resistance genes in aquatic systems at low-level oxytet- racycline exposures. Environ Sci Technol 42:5348–5353

18. Kümmerer K (2001) Pharmaceuticals in the environment: sources, fate, effects and risks. Springer, Berlin

19. Laurent J, Casellas M, Dagot C (2009) Heavy metals uptake by sonicated activated sludge: relation with floc surface properties. J Hazard Mater 162:652–660

20. Li B, Zhang T, Xu ZY, Fang HHP (2009) Rapid analysis of 21 antibiotics of multiple classes in municipal wastewater using ultra performance liquid chromatography-tandem mass spectrometry. Anal Chim Acta 645:64–72

21. Liu XM, Sheng GP, Luo HW, Zhang F, Yuan SJ, Xu J, Zeng RJ, Wu JG, Yu HQ (2010) Contribution of extracellular polymeric substances (EPS) to the sludge aggregation. Environ Sci Technol 44:4355–4360

22. Martínez JL (2008) Antibiotics and antibiotic resistance genes in natural environments. Science 321:365–367

23. Miao XS, Bishay F, Chen M, Metcalfe CD (2004) Occurrence of antimicrobials in the final effluents of wastewater treatment plants in Canada. Environ Sci Technol 38:3533–3541

24. Monteiro SC, Boxall ABA (2009) Factors affecting the degradation of pharmaceuticals in agricultural soils. Environ Toxicol Chem 28:2546–2554
25. Pérez S, Eichhorn P, Aga DS (2005) Evaluating the biodegradability of sulfamethazine, sulfamethoxazole, sulfathiazole, and trimethoprim at different stages of sewage treatment. Environ Toxicol Chem 24:1361–1367
26. Pevere A, Guibaud G, van Hullebusch ED, Boughzala W, Lens PNL (2007) Effect of Na$^+$ and Ca^{2+} on the aggregation properties of sieved anaerobic granular sludge. Colloids Surf A 306:142–149
27. Pils JRV, Laird DA (2007) Sorption of tetracycline and chlortetracycline on K- and Ca-saturated soil clays, humic substances, and clay-humic complexes. Environ Sci Technol 41:1928–1933
28. Sarmah AK, Meyer MT, Boxall ABA (2006) A global perspective on the use, sales, exposure pathways, occurrence, fate and effects of veterinary antibiotics (VAs) in the environment. Chemosphere 65:725–759
29. Sassman SA, Lee LS (2005) Sorption of three tetracyclines by several soils: assessing the role of pH and cation exchange. Environ Sci Technol 39:7452–7459
30. Smith PG, Coackley PA (1983) Method for determining specific surface area of activated sludge by dye adsorption. Water Res 17:595–598
31. Sørensen BL, Wakeman RJ (1996) Filtration characterisation and specific surface area measurement of activated sludge by rhodamine B adsorption. Water Res 30:115–121
32. Spongberg AL, Witter JD (2008) Pharmaceutical compounds in the wastewater process stream in Northwest Ohio. Sci Total Environ 397:148–157
33. Sun XF, Wang SG, Zhang XM, Chen JP, Li XM, Gao BY, Ma Y (2009) Spectroscopic study of Zn^{2+} and Co^{2+} binding to extracellular polymeric substances (EPS) from aerobic granules. J Colloid Interface Sci 335:11–17
34. Sun HY, Shi X, Mao JD, Zhu DQ (2010) Tetracycline sorption to coal and soil humic acids: an examination of humic structural heterogeneity. Environ Toxicol Chem 29:1934–1942
35. Tanis E, Hanna K, Emmanuel E (2008) Experimental and modeling studies of sorption of tetracycline onto iron oxides-coated quartz. Colloids Surf A 327:57–63
36. Ter Laak TL, Gebbink WA, Tolls J (2006) The effect of pH and ionic strength on the sorption of sulfachloropyridazine, tylosin, and oxytetracycline to soil. Environ Toxicol Chem 25:904–911
37. Wan Y, Bao YY, Zhou QX (2010) Simultaneous adsorption and desorption of cadmium and tetracycline on cinnamon soil. Chemosphere 80:807–812
38. Wang YJ, Jia DA, Sun RJ, Zhu HW, Zhou DM (2008) Adsorption and cosorption of tetracycline and copper(II) on montmorillonite as affected by solution pH. Environ Sci Technol 42:3254–3259
39. Wilén BM, Jin B, Lant P (2003) The influence of key chemical constituents in activated sludge on surface and flocculating properties. Water Res 37:2127–2139
40. Xiao Y, Chang H, Jia A, Hu JY (2008) Trace analysis of quinolone and fluoroquinolone antibiotics from wastewaters by liquid chromatography–electrospray tandem mass spectrometry. J Chromatogr A 1214:100–108
41. Xu XR, Li XY (2010) Sorption and desorption of antibiotic tetracycline on marine sediments. Chemosphere 78:430–436

42. Xu ZY, Fan J, Zheng SR, Ma FF, Yin DQ (2009) On the adsorption of tetracycline by calcined magnesium-aluminum hydrotalcites. J Environ Qual 38:1302–1310

43. Zhang LL, Zhu DQ, Wang H, Hou L, Chen W (2012) Humic acid- mediated transport of tetracycline and pyrene in saturated porous media. Environ Toxicol Chem 31:534–541

There are several supplemental files that are not available in this version of the article. To view this additional information, please use the citation on the first page of this chapter.

PART II

DIVERSITY OF ACTIVATED SLUDGE AND PROCESS MICROBIOLOGY

CHAPTER 5

Evaluation of Simultaneous Nutrient and COD Removal with Polyhydroxybutyrate (PHB) Accumulation Using Mixed Microbial Consortia under Anoxic Condition and Their Bioinformatics Analysis

JYOTSNARANI JENA, RAVINDRA KUMAR, ANSHUMAN DIXIT, SONY PANDEY, AND TRUPTI DAS

5.1 INTRODUCTION

Polyhydroxyalkanoate (PHA), a bio-polymer currently under scrutiny as an alternative to petroleum based plastics, is a metabolic by-product of microorganisms active in wastewater treatment plants (WWTP) operating under high substrate load [1–6]. Wallen and Rohwedder [7] for the very first time reported PHA accumulation in microorganisms performing Enhanced Biological Phosphate Removal (EBPR) in a WWTP. Now it is well known that a conventional EBPR process achieves nutrient removal

Evaluation of Simultaneous Nutrient and COD Removal with Polyhydroxybutyrate (PHB) Accumulation Using Mixed Microbial Consortia under Anoxic Condition and Their Bioinformatics Analysis. © Jena J, Kumar R, Dixit A, Pandey S, Das T. PLoS ONE 10,2 (2015). doi:10.1371/journal. pone.0116230. Licensed under a Creative Commons Attribution 4.0 International License, http://creativecommons.org/licenses/by/4.0/.

under a sequential anaerobic-aerobic/anoxic state by a specific group of microorganisms termed as PAOs (Phosphate Accumulating Organisms). Depending on the substrate type, either polyhydoxybutyrate (PHB) or polyhydroxyvalerate (PHV) is synthesized by the microorganisms in the anaerobic phase along with phosphate release due to splitting of the internal polyphosphate pool to produce the required ATP [8]. In the subsequent electron rich aerobic/anoxic phase, stored PHB is utilized to replenish the phosphate pool resulting in overall phosphate removal [8–10]. Therefore, it is obvious that simultaneous phosphate accumulation and PHB recovery is not a feasible option in a conventional EBPR system. PHB recovery from EBPR enriched sludge is a well documented phenomenon, however, most of these studies don't focus on nutrient removal [11, 12]. Reports on EBPR system achieving PHB accumulation without disturbing nutrient removal are sparse [12, 13].

Apart from operating conditions, a range of diverse micro flora might also contribute to simultaneous PHB accumulation and nutrient removal. Initially the presence of nitrate was believed to have an inhibitory effect on phosphate accumulation by PAOs, as nitrate rich medium was observed to be more suitable for proliferation of the denitrifiers [1]. Later it was observed that under anaerobic/anoxic system phosphate accumulation was being performed by a group of denitrifiers, using the intracellular PHA/PHB as electron donor and nitrate as terminal electron accepter. Subsequently this group of microorganisms was termed as DNPAOs [8]. Schuler and Jenkins [13] and later Zhou et al. [14] reported that by means of glycogen degradation as an alternate energy source, PAOs can accumulate PHA without releasing phosphate. A few other reports also suggest that denitrifying bacteria don't have to undergo PHB hydrolysis for phosphate uptake [15, 16]. Efficacy of the dynamic EBPR sludge has been validated for PHB recovery under different operational conditions like complete aerobic, aerobic dynamic feeding with variables like HRT, pH, SRT etc [17, 18]. However, there is a dearth of information as to what would be the behaviour of the microorganisms and the reactor performance under complete anoxic conditions with surplus external substrate. Under such a state, when both electron donor and acceptor are available in surplus, the microorganisms might function differently, which in turn might alter the scenario of conventional EBPR.

The anoxic system, if optimized, can find a way to resolve two major environmental challenges of the current era, i.e, industrial/municipal effluent treatment along with the production of a biodegradable compound. A carbon source in anoxic condition has been reported to inhibit phosphate uptake [19, 20]. At a later stage, researchers have determined a specific group of bacteria capable of accumulating phosphate in the presence of both nitrate and acetate [15]. Substrate (carbon) concentration is understood to be analogous with high PHB accumulation, however nutrient release has also been reported under similar conditions [3, 6]. Therefore, it is crucial to optimize the carbon concentrations in the reactor en route for development of a sustainable process that can strike a balance between optimum nutrient removal with simultaneous PHB accumulation.

In the current work, a set of batch experiments has been designed to study the potential of enriched microbial consortia for simultaneous nutrient removal and PHB accumulation under complete anoxic conditions with varying ICL. As a mixed bacterial consortium help maintaining the dynamism of the overall system; therefore the microbial analysis of the seed sludge was conducted using high throughput sequencing techniques, to determine the key players in the process.

5.2 MATERIALS AND METHODS

5.2.1 SYNTHETIC WASTE WATER

Synthetic wastewater (KNO_3–1.63 g/L as nitrate source, KH_2PO_4–0.043 g/L as phosphate source, $MgSO_4$–1.5 g/L, peptone-0.38 g/L), and 0.3mL of nutrient solution (0.15g/L $FeCl_3.6H_2O$, 0.15 g/L H_3BO_3, 0.03g/L $CuSO_4.5H_2O$, 0.18g/L KI, 0.12 g/L $MnCl_2.4H_2O$, 0.06 g/L $Na_2MoO_4.2H_2O$, 0.12 g/L $ZnSO_4.7H_2O$, 0.15g/L $CoCl_2.H_2O$, 10g/L EDTA) was used in all the batch reactors [21]. Synthetic waste water was sterilized prior to the experiment. Sodium acetate was added in different concentrations (2 g/L-ICL1; 4 g/L-ICL2; 6 g/L-ICL3; 8 g/L-ICL4 and 10 g/L-ICL5) in order to vary the ICL and depending on the concentration of carbon, the initial COD/NO_3 ratio in different batch experiments were 1.75 for ICL-1, 3.7 for ICL-2, 5.3 for ICL-3, 7.1 for ICl-4, 8.9 for ICl-5. Deliberately high

COD/NO$_3$ ratio was opted to evaluate the reactor performance to achieve simultaneous COD and nutrient removal along with PHB accumulation by the microbial consortia.

5.2.2 BATCH REACTOR SET UP

Reactors having 1L of working volume connected with a pH probe and nitrogen purging tube were used as the batch reactors for each set of experiment. 50 mL of activated sludge from a Sequencing Batch Reactor (SBR) performing biological nutrient removal was used as inoculum in each set. After inoculation, the reactors were sealed and aseptic conditions were maintained in order to prevent any contamination during the experimental period.

In the batch reactors nitrogen gas (~99.99% purity) was passed throughout the run time to maintain complete anoxic condition. The initial pH of the medium ranged between 7.2–7.5. Representative samples were drawn for the analysis of various parameters at regular intervals within the experimental run time of 6 h.

To validate the exact effect of nitrate on phosphate and COD removal, one set (ICL2) was repeated with longer experimental period. The reactor was supplemented with 0.5g/L of nitrate solution at the 6th hour of the experiment, i.e. after exhaustion of initial nitrate. Then the reactor performance was monitored for the next 6 hours (depending on the time taken for complete exhaustion of nutrient and the residual COD).

5.2.3 OPERATIONAL CONDITION OF SBR PROVIDING SEED SLUDGE FOR THE BATCH STUDIES

Activated sludge was drawn at the beginning of anoxic phase from the SBR and was inoculated to each set of batch reactor as seed sludge. This was done to maintain the uniformity of microbial population during the batch experiments.

Composition of synthetic wastewater and all the other parameters like initial pH, NO$_3$, PO$_4$, COD concentrations and time period of the anoxic cycle of SBR was similar to the batch studies excepting the ICL. The SBR was continuously being operated for 180 days in a 12 hour cycle with anoxic (6h) -aerobic (5h)—settle-decant-refill (1h) phases. It was validated for high nutrient removal efficiency during each cycle, which supports the prevailing steady state condition during the period of operation of the SBR (unpublished data). However the potential of microorganisms (in the SBR) for PHB accumulation was only validated with sudan black staining prior to the batch experiments.

5.2.4 ANALYTICAL METHODS

Collected samples were filtered (0.45 mm pore size), filtrates were analyzed for nitrate-N, phosphate and COD using standard techniques [22]. Volatile Suspended Solid (VSS) measurement was done for each experimental set up. Sudan black staining procedure was used to substantiate the accumulation of PHB granules inside the microbial cells. Extraction and estimation of PHB were performed using standard protocol [3, 23].

5.2.5 CHARACTERIZATION OF PHB

5.2.5.1 SOLID STATE H-NMR ANALYSIS

Chemical analysis of the polymer was done using 1H nuclear magnetic resonance (NMR). 1 mg of the polymer was dissolved in 1 mL of $CDCL_3$ and the spectra of the standard (commercially available PHB from SIGMA Aldrich) as well as test samples were obtained by using a Jeol FT-NMR (400MHz of the coupling constant).

5.2.5.2 FOURIER TRANSFORMS INFRARED SPECTROSCOPY (FTIR)

The extracted PHB was further analyzed by FTIR using available standards as reference. A PerkinElmer spectrum GX FTIR was used for analysis with a spectral range of 4000 cm^{-1} to 40 cm^{-1} and resolution of 0.15 cm^{-1}.

5.2.6 MICROBIAL COMMUNITY EXPLORATION

Microbial community was studied by 16s rDNA analysis. DNA samples extracted from the seed sludge during the beginning of anoxic phase of SBR were analyzed through pyrosequencing technique. This sludge was used in each batch experiment conducted under aseptic conditions. It is quite evident that though there may be a variation in the population size in different batch studies, the overall microbial community is supposed to be the same as in the SBR.

In a separate SBR which was operational under similar conditions as mentioned above; 16S rDNA analysis of the microbial community revealed a similarity in population dynamics with change in operational conditions (Unpublished data). This again suggests that the population dynamics in the batch reactors would have been similar to the dynamics in the seed sludge.

Therefore, batch reactor performance was supported by both chemical and microbial community (16S rDNA) analysis rather than solely depending upon any one technique.

Total DNA was isolated from the seed sludge sample using a DNA isolation kit (Fast DNA Spin Kit, MP Biomedicals) as per the manufacturer's protocol. Bacterial 16S rDNA genes from the total bacterial DNA were amplified by PCR followed by pyrosequencing at the Research and Testing Laboratory (RTL), Texas, USA.

In the current study, we have used two different software programs, namely Ribosomal database project (RDP) [24] and DECIPHER [25]. The RDP is a collection of various tools to analyze data for microbial population from large sequencing libraries. DECIPHER has extensive capabilities for identification of chimeric sequences. The analysis of sequencing data was done as outlined below.

5.2.6.1 PRE-PROCESSING

Raw reads were treated by pyrosequencing pipeline of RDP modules using default settings. Primers, adaptors and barcodes were trimmed from each read. The sequence reads with poor quality scores were removed from further analysis. Thereafter, DECIPHER was used to filter chimeric sequences. The remaining sequences reads were used for further analysis.

5.2.6.2 CALCULATION OF RAREFACTION CURVE AND DIVERSITY INDEXES

The pre-processed sequence reads were aligned by infernal [26] using the bacteria-alignment model (model-9). Complete linkage clustering was performed to assign these reads to the Operational Taxonomic Units (OTUs) at the distance threshold of 0.01, 0.03, 0.05, 0.07 and 0.10. Rarefaction curve and diversity indices (Chao1 richness and Shannon index) were calculated on the basis of these clusters. These steps were done using RDP.

5.2.6.3 TAXONOMIC CLASSIFICATION

The sequence reads were classified into different taxonomic classes using RDP Classifier tool. It uses a naïve Bayesian classification algorithm for classification of sequences into different taxonomic classes. A bootstrap cutoff of 50% as suggested by RDP developers was applied to assign the sequences to different taxonomic levels.

5.3 RESULT AND DISCUSSION

5.3.1 NUTRIENT AND COD REMOVAL

Nutrient removal efficiency has been evaluated by monitoring the concentration of phosphate and nitrate-N in each set of experiment for a period of 6 h. Beyond 6 hours nutrient removal was depleted as more than 95% of nitrate-N was utilized. Hence the reactor run time was kept limited to 6 h initially. Fig. 1 represents the percentage removal of phosphate, nitrate-N and COD under different ICL. It has been assumed that the COD load in the batch reactor is directly proportional to the ICL [27]. It can be observed in Fig. 2 (a) that variation of ICL had a direct impact on nitrate-N removal rate.

Within 6 hours of incubation nitrate-N removal was significant in ICL-1 (81%), ICL-2 (92.3%), ICL-3 (96.4%), ICL-4 (97.9%) and ICL-5 (84.7%). Low initial carbon load (in ICL-1) might have hindered the nitrate-N utilization rate [28]. ICL-2, 3 and 4 showed a substantial and sequential rise in the nitrate removal rate indicating the importance of an optimal COD/NO_3 ratio in the medium that supports the growth of the enriched biomass playing a major role in the nutrient removal process. These results are concordant with previous studies [27, 29, 30]. In ICL-5 nitrate removal rate declines in spite of a high COD/NO_3 ratio, which has been reported earlier to have favoured nitrate reduction [30, 27]. In ICL-5 a rise in carbon concentration also lowers the phosphate/COD ratio to 0.004 in the medium which has been stated [31, 32] to be a favourable condition for the metabolism of other heterotrophic organisms (OHOs) like Glycogen Accumulating Organisms (GAOs) [33–35]. Under similar conditions PAOs are also reported to behave as GAOs [14] and a low phosphate/COD ratio is usually preferred for enrichment of GAOs [31, 36]. GAOs are capable of nitrate reduction and acetate uptake under anoxic condition and are termed as DGAOs (Denitrifying Glycogen Accumulating Organisms), however, they lack the ability to recycle phosphate [37, 38]. Therefore, it is evident that in ICL-5, DNPAO performances were partially affected by other heterotrophs like DGAOs which could have become more active due to suitable growth conditions. This is in agreement with other studies [14, 31, 37]. Microbial analysis of the seed sludge (described in a later section) has revealed the co-existence of DNPAOs and DGAOs in the mixed consortium however, due to the prevailing conditions in ICL-5 the population size of both DNPAOs and DGAOs might have been altered during the course of the experiment.

It was observed that variation in ICL also had a significant impact on the phosphate removal trend (Fig. 1, Fig. 2b). Overall, 81.5%, 82.4% and 87.4% of PO_4 removal was achieved in ICL-2, 3 and 4 respectively within 6 hours of incubation. Discrepancy in PO_4 removal in ICL-1 (49.3%), and ICL-5 (42.8%) might be due to scarcity of carbon source in the former and high concentration of carbon that disturbs the growth and maintenance of

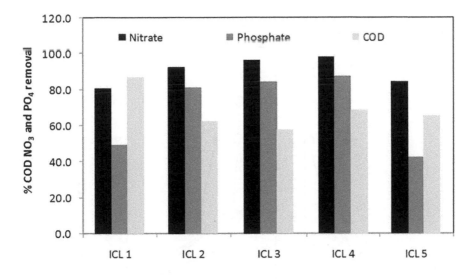

Figure 1: % of removal of phosphate, nitrate-N and COD under different ICL.

DNPAOs in the later as described in the earlier section [27, 39]. A clear explanation can be derived by comparing Fig. 2a with Fig. 2b. In ICL-2 and 3, nitrate-N concentration touches the minimum value at 5 hours of incubation, and the PO_4 concentration goes up (Fig. 2b) which indicates that a decline in nitrate-N concentration in the medium puts stress on the microorganisms and probably deplete the stored polyphosphate for energy; thus releasing inorganic PO_4 to the medium. Further, it is observed that in ICL-4, PO_4 release occurs beyond 4 hours of incubation with the subsequent drop in nitrate-N concentration almost during the same time. Interestingly, in ICL-1 and ICL-5, PO_4 was released into the medium beyond 2 hours of incubation, though during that time nitrate-N concentration was plenty. The carbon load in ICL-1(less than optimal) and ICL-5 (more than optimal) might have triggered a stressful state for DNPAOs leading to an early PO_4 release by the microorganisms as described earlier. Under the prevailing conditions in ICL-5, DGAOs/OHOs might have brought about a partial disturbance in phosphate utilization by DNPAOs, limiting the

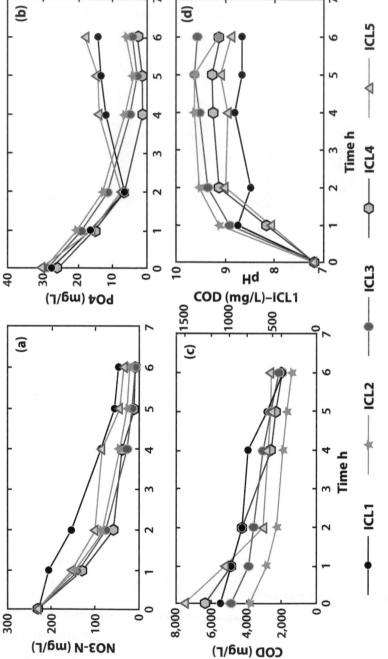

Figure 2: NO$_3$ (a), PO$_4$ (b), COD (c) removal and pH (d) trend in the batch reactors operated for 6 h.

phosphate uptake to only 42.8%. These observations are at par with previous results [35, 40].

Highest COD removal was achieved in ICL-1 (87%) followed by ICL-4 (69%), ICL-5 (66%), ICL-2 (63%) and ICL-3 (58%) (Fig. 2c). Initial nitrate-N concentration in ICL-1 was sufficient to facilitate maximum utilization of minutely available COD by the biomass. In ICL- 5 considerable COD and nitrate-N removal with phosphate release was observed beyond 2 hours of incubation. Previous studies have also reported the outgrowth of DNPAOs by OHOs under a high COD/NO_3 ratio as the nitrate load is insufficient to support the process of denitrification by DNPAOs [35, 41]. Therefore, in ICL-5 a high COD/NO_3 ratio along with a low phosphate/COD ratio supports proliferation of OHOs/DGAOs. In other sets (ICL-2, 3 and 4) a rapid consumption of nitrate-N left the amount of residual COD high. It has been explained later that the addition of extra nitrate-N to the reactor after exhaustion (beyond 6h) brought about a remarkable decrease in the final COD (86% of removal was achieved in ICL-2).

5.3.1.1 IMPORTANCE OF NITRATE-N IN AN ANOXIC REACTOR

The efficiency of all batch experiments, described in the prior section, was disturbed following nitrate-N exhaustion in the medium. Therefore, it was apparent that nitrate-N was the limiting factor and depletion of nutrient removal rate was directly correlated with availability of electron acceptor [42, 43] i.e. nitrate-N in the present study. Exhaustion in nitrate-N concentration generates stress that probably leads to breakdown of Poly P and release of phosphate to the medium [44]. Further investigations were carried out by supplying extra nitrate-N (110 mg/L) into the anoxic reactor followed by the exhaustion of initial nitrate-N supplied at the beginning of the reaction. It was observed that with addition of extra nitrate-N, to the batch-reactor (ICL-2) in the 7th hour of experiment, more than 90% of PO_4 (Fig. 3a) and COD removal was achieved (Fig. 3b).

Results of the batch studies indicate that concentration of initial nitrate-N and carbon load are interdependent parameters for achieving ideal nutrient and COD uptake from wastewater. It was also presumed that in spite of a high carbon load, the performance of an anoxic reactor can be maintained at an ideal state (achieving maximum COD and nutrient removal) provided adequate nitrate-N is available for utilization as electron acceptor by the microorganisms. From this set of experiment COD/NO_3 ratio of 2.3 was found to be optimum for the growth and metabolism of the microbial consortium in the present case to achieve maximum nutrient and COD removal.

5.3.2 PH TREND IN THE EXPERIMENT

Irrespective of initial carbon load, pH of the reaction medium showed a rising trend in all the sets due to prevailing denitrifying condition (Fig. 2d) [45]. Increase in pH along with nitrate-N reduction was a virtual indicator of active metabolic trait of DNPAOs in the reactor. The increasing trend of pH was seized after exhaustion of nitrate-N (6h), indicating a decline in the rate of denitrification. Further it can be noticed that the pH trend is alike in ICL-2, ICL-3 and ICL-4 for which nitrate-N removal rate is more or less similar. Whilst in ICL-1 the trend shows an initial increase followed by a gradual downfall beyond 3 h of incubation and in ICL-5 the trend is slightly uneven. It indicates the discrepancies in denitrification rate in both the sets which ultimately results low in nutrient removal.

5.3.3 KINETIC STUDY

In this study nitrate-N was found to be the limiting factor for nutrient and COD removal. Therefore Rate of nitrate-N utilization under various ICL can be explained by the following equation:

$$Rate = \frac{-dN}{dt} = K(N)$$

Where N is the nitrate-N and K is the specific reaction rate constant. Integration of (eq 1) gives the following

$$-\ln \frac{N_i}{N_0} = Kt$$

Where N_t is the concentration of nitrate-N at time t and N_0 is the initial nitrate-N concentration in the reaction medium.

Considering the reaction to be first order a plot of ln N_t versus time gives a straight line and N_0 from the slope k was calculated (Fig. 4). The R-squared value for each line was observed to be >0.9. Hence it was assumed that the nitrate-N removal rate by activated sludge was a first order reaction. Values of 'k' are 0.311, 0.419, 0.552, 0.63 and 0.283 for ICL-1, ICL-2, ICL-3, ICL-4 and ICL-5 respectively.

5.3.4 PHB ACCUMULATION

Sudan black staining of biomass confirmed the accumulation of PHB [3]. DNPAOs uptake excess external carbon and store it as PHB in the presence of nitrate-N as luxury PHB uptake [2]. It was further observed that nitrate-N did not have an adverse effect on PHB accumulation in the current study, as was reported earlier [1, 46].

Fig. 5 shows the variability of PHB accumulation in the bacterial cell (mg of PHB accumulated per 1L activated sludge) with respect to different ICL. It can be observed that PHB accumulation was directly proportional to the organic carbon load; with less PHB accumulation at lower carbon load and vice versa (except ICL5). However, time taken for high PHB accumulation (5h in ICL-2 and beyond 5h in ICL-3) increased with an increase in initial carbon concentration [46]. In ICL-5, the PHB accumulation shows an increasing trend but during 6 hours of incubation 30 mg/L of PHB was accumulated which was lower than ICL-3 and 4, where more than 45 mg/L of PHB accumulation occurred during the same time period. As has been described earlier DNPAO population, supposed to accumulate PHA are impaired which could have lead to the discrepancy in PHB accumulation observed in ICL-5.

ICL-1 experienced a slight increase in PHB concentration, due to utilization of ~200mg/L of COD within 2 hours of incubation and a slight fall in concentration was observed beyond that; even though the COD concentration was showing a decreasing trend during the same time. This inconsistency might be caused due to a lower concentration of external carbon. It is obvious that PHB accumulation occurs only when the availability

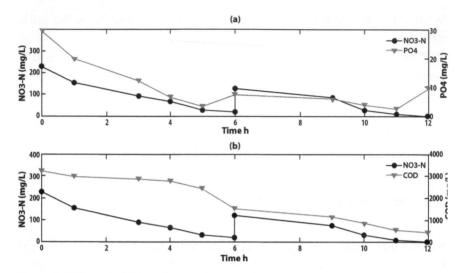

Figure 3: Effect of addition of extra nitrate-N after exhaustion of initial nitrate-N in the reactor on (a) phosphate removal (b) COD removal.

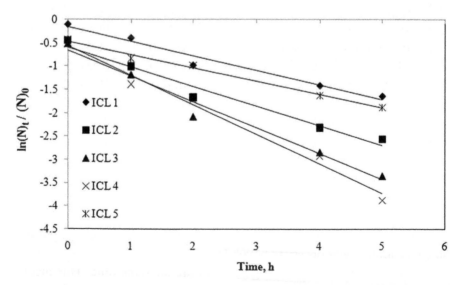

Figure 4: First order plot for nitrate-N utilization under various initial carbon load (ICL-1 to ICL-5).

of external carbon source is in excess [47, 48] but interestingly a decline in PHB concentration is also marked with simultaneous release in PO_4 concentration indicating a stressful environment for DNPAO population. This typical observation further justifies the fact that a part of COD might have been utilized for growth and maintenance of the bacterial consortia. However, the exact reason behind the PHB break down with simultaneous phosphate release beyond 2 hours of incubation could not be addressed in the current study. Could this be a mechanism adapted by the consortia to deal with stress conditions? Further investigation is required to validate such an assumption.

Similarly the slight depletion in the PHB along with phosphate release beyond 5 h of incubation in ICL-2 indicates the development of a stress condition followed by nitrate-N exhaustion as described earlier. In conventional EBPR, hydrolysis of PHB facilitates poly P build up [49, 50] but this hypothesis could not be justified in the current study as PHB depletion was not accompanied by subsequent phosphate uptake from the medium rather phosphate was released beyond 5 h of incubation (Fig. 2). 16s rDNA analysis of the seed sludge is evident of the fact that the most abundant population in the sludge, Paracocous does not utilize PHB for Poly P formation [15].

5.3.5 CHARACTERIZATION OF THE POLYMER

5.3.5.1 H-NMR ANALYSIS

H-NMR analysis was used to determine the structure of the polymer recovered from the sludge. The spectrum confirmed the polymer to be PHB showing a doublet peak at 1.25 ppm alkyl (secondary and tertiary), doublet of quadruplet peaks between 2.4ppm attributed to a methylene group adjacent to asymmetric carbon atom and a single peak at 5.2ppm also attributed to a methylene group (Fig. 6) [51, 52]. A large peak in between 7–7.5ppm might be due to the departed CDCL3 and the peak at 0.01 is due to water contamination.

5.3.5.2 FUNCTIONAL GROUP ANALYSIS USING FTIR

PHB crystals were extracted and analyzed through FTIR. The spectrum of the extracted polymer is shown in Fig. 7. The band found at 1459 cm^{-1} corresponded to the asymmetrical C-H bending vibration in the -CH$_3$ group while the band found at 1120 cm^{-1} was equivalent to >CH$_2$ symmetrical bending vibration. The strong absorption band at 1733 cm^{-1} indicated stretching of the >C = O bond and the band at 1280 cm^{-1} corresponded to the -CH group. The series of bands located in the region of 1000 to 1200 cm^{-1} corresponded to the stretching of the >C-O bond of the ester group. The absorption band around 3000 cm^{-1} was assigned to the terminal -OH group. The absorption peaks obtained are comparable with earlier reports [52, 53] and with the spectrum of pure PHB standard, thus confirming the extracted polymer to be PHB.

FTIR analysis was also used to measure the degree of crystallinity of the polymer. The crystallinity index (CI) is defined as the ratio of the intensity of the bands at 1382 cm^{-1} (CH$_3$), which are insensitive to the degree of crystallinity, to that at 1185cm^{-1} (>C-O-C<), sensitive to an amorphous state. CI of the extracted PHB was found to be 0.86, which was lower than a previously reported value of 0.949 [53]. As lower crystallinity induces a faster degradation of the polymer, it is expected to play a key role in determining the polymer stability.

5.3.6 MICROBIAL COMMUNITY ANALYSIS

Total 11252 sequence reads were acquired from 454 sequencing of the microbial community 16S rDNA, after initial filtering 9683 reads were finally selected for further investigation. Detailed results of preprocessing are given in Table 1.

5.3.6.1 RAREFACTION CURVE ANALYSIS

Rarefaction analysis was employed to analyze taxon richness of the sample and to check adequacy of sampling. Plot of the number of sequences

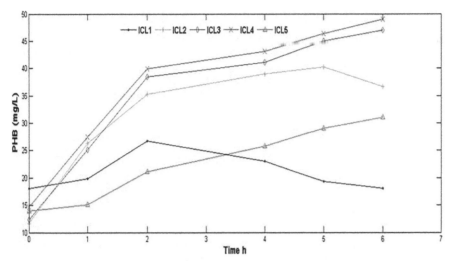

Figure 5: PHB accumulation in different sets with different ICL.

against number of OTUs is called rarefaction curve. Distance cut-off values of 0.03, 0.05, 0.07 and 0.10 are generally accepted as points at which differentiation occurs at the species, genus and family/class level, respectively [54]. The upward slope on the curve reflects that a large number of species remain to be discovered while flatter line indicates that almost all the species in the sample have been discovered. It also indicates about the adequacy of the sampling required to discover total species diversity in that ecosystem. Rarefaction curve in Fig. 8 at the cut-off level of 1%, 3%, 5%, 7% and 10% show that slopes tend to be flat, which means the sequence analysis represents almost complete bacterial population and corresponding number of OTUs.

5.3.6.2 CALCULATION OF DIVERSITY INDEXES

The diversity of sampling was analyzed using Chao1 and Shannon index (H') (Table 2). "UCI95" and "LCI95" indicate 95% confidence level of upper and lower values of Chao1 for each distance.

It can be seen in table 2 that the values of Chao1 index also support the results of rarefaction curve. The evenness "E" refers to how close in numbers each species is in an environment. The H' value range (1.84–3.09) and species evenness (E= ~0.5) indicate moderate species diversity. This is comprehensible as the current conditions are suitable to augment a specific array of microbial consortia to persistently achieve nutrient removal along with PHB accumulation.

5.3.6.3 TAXONOMIC ANALYSIS

Reads were classified into different taxa levels (from phylum to genus) using RDP classifier at 50% thresh-hold as per the recommendation of Cole et al. [24]. The results are highlighted in Fig. 9(A–E).

Details of analysis of the sludge used in the batch experiments have been provided in Tables S1, S2, S3, S4 and S5 in S1 File. When these results were compared with 16S rDNA analysis of sludge from a separate SBR (as mentioned earlier), much similarity was observed in the population dynamics between both the microbial communities (Table S1a, S2a, S3a, S4a and S5a in S1 File) with *Proteobacteria, Alaphaproteobacteria, Rhodobacterales, Rhodobacterececae* and *Paracoccus* being the dominant phylum, class, order, family and genus respectively. These observations further reveal that with change in operational conditions variation in population size were observed in both the sludges, whereas the overall microbial communities were highly similar. The results also show a continuous increase in the unclassified sequences from phylum to family (from 3.53% to 13.61%). It indicates that at the phylum level most of the sequences were classified but at the family and genus level they were not completely categorized. It may also indicate the presence of some novel families or genus of bacteria [54].

Total six bacterial phylum were assigned and *Proteobacteria* (69.23%), *Actinobacteria* (8.35%) and TM7 (3.53%) were found to be predominant among them (Fig. 9A).

Proteobacteria have been frequently reported to be the most prevailing species in biological waste water treatment plants (WWTP) [54–56]. *Actinobactors* are the filamentous bacteria largely reported in WWTPS performing EBPR [55]. Some also suggest it's availability in Denitrifying Phosphorous Accumulating processes [57].

Presence of metabolically active TM7 has been observed in activated sludge as well as in WWTPs however their exact function is yet to be elucidated. Researchers are also inquisitive to establish a precise link between TM7and human health as the former has been detected in skin and oral samples of human beings along with its presence in environmental sites [55, 58].

Main classes in the sample were *Alphaproteobacteria* (69%), *Betaproteobacteria* (12%), *Actinobacteria* (8%) and *Gammaproteobacteria* (2%) (Fig. 9B). However this observation was different from previous studies which showed that *Gammaproteobacteria* and *Betaproteobacteria* were predominant in sludge sample [54, 59].

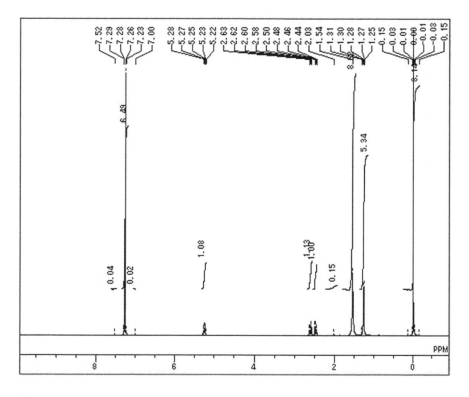

Figure 6: H-NMR spectra of extracted Polymer.

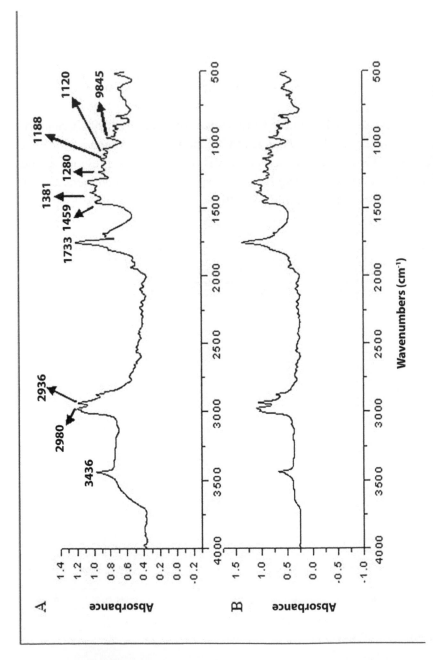

Figure 7: FTIR spectra of standard poly (3–hydroxybutyric acid) (B) and the polymer isolated from the reactor (A).

Table 1. Quality filters results using RDP pyro sequencing pipeline.

Initial reads	Exponential quality filter	Avg. length after trimming	S.D. of length	Chimeric Sequences	Reads after pre processing
11252	949	384	45.55	620	9683

This difference may be attributed to the reactor conditions as it is reported that *Alphaproteobacteria* class performs denitrification and PHB accumulation in anoxic condition under dynamic substrate availability [60] which was further supported by detection of *Paracoccus* (class *Alphaproteobacteria*) as the dominant genus in the sludge sample (Fig. 9E).

Twelve orders were found in the total bacterial population of the sludge sample. The most dominant orders were *Rhodobacterales* (69.23%), *Rhodocyclales* (9.67%), *Actinomycetales* (8.35%), *Burkholderiales* (1.98%) and *Pseudomonadales* (1.14%) (Fig. 9D).

Eighteen families and 27 genus were detected in the sludge sample. The dominant genus was *Paracoccus* (6695 sequences), *Azoarcus* (684 sequences), *TM7_genera_incertae_sedis* (342 sequences), *Brachymonas* (143 sequences), *Corynebacterium* (140 sequences) and *Azonexus* (138 sequences) which accounted for 69%, 7%, 4%, 1%, 1% and 1% respectively of the total sequence reads. The genus could not be classified for 12.5% sequences. The most prevalent families were *Rhodobacteraceae* (69%), *Rhodocyclaceae* (10%) *Comamonadaceae* (2%), *Corynebacteriaceae* (1%), *Pseudomonadaceae* (1%). Family level classification for 14% sequences could not be done (Fig. 9C).

The most abundant species in this sludge, *Paracocous* (Fig. 9E), is capable of simultaneous phosphate and nitrate removal in wastewater treatment system under anoxic conditions. This species has also been widely reported to uptake short chain fatty acid and store it as polyhydroxybutyrate (PHB) in both WWTP and single culture system [15, 61]. However these microorganisms are unable to utilize PHB for Polyphosphate build up [8, 15] which is also evident in the current study. As the microorganisms tend to show a different behaviour under various growth conditions, therefore the specific mechanism behind substrate utilization and growth is also subject to change under various operational parameters [8, 15, 62–67]. *Azoarocous, Azonexus* and *Thaurea*, member of family *Rhodocyclaceae* are reported to be involved in phosphate accumulation [55]. *Corynebacterim* species are well documented as phosphate accumulators in WWTPs [55, 56, 68]. It has been reported that PAOs have the ability to utilize both oxygen and nitrate as electron acceptor [14] therefore in the present case, specific microorganisms might have utilized nitrate as an electron acceptor to accumulate phosphate. Among other sparsely available bacterial population, *Pseudomonas* species have been reported to perform heterotrophic denitrification along with PHB accumulation [69, 70]. *Gammaproteobacteria* widely reported as GAOs [18, 32], *Firmicutes* and other 12.5% of the unclassified bacterial population could be the OHOs/

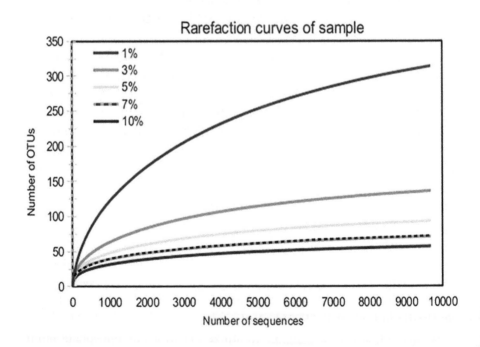

Figure 8: Rarefaction curve of sludge sample at cutoff level of 1%, 3%, 5%, 7% and 10%.

Table 2: The number of OTUs, Chao1 and Shanon index (H').

Level	Clusters	Chao1	LCI95	UCI95	H'	Var H	E
1%	314	360.7747	340.7526	395.7816	3.09577	0.00049	0.53845
3%	136	152.24	142.5113	176.5045	2.52768	0.00033	0.51452
5%	93	101	95.47581	118.8502	2.18648	0.00031	0.48239
7%	71	76.5	72.40662	92.50543	2.11933	0.00028	0.49718
10%	57	63.42857	58.51363	84.3029	1.84166	1.84166	0.45551

GAOs which attribute to the performance variation of the batch reactors. Though TM7 is one of the most populated groups in the consortium, its exact function is yet to be elucidated.

5.4 CONCLUSIONS

Batch studies performed with enriched microbial consortia and varying ICL, under complete anoxic conditions exhibited simultaneous nutrient and COD removal along with PHB accumulation. Within 6 hours of incubation, enriched DNPAOs were capable of achieving maximum COD (87%) removal at 2g/L of ICL whereas maximum nitrate-N (97%) and phosphate (87%) removal along with PHB accumulation (49 mg/L) was achieved at 8 g/L of ICL. Being an anoxic set up, nitrate-N concentration was observed to be the limiting factor as exhaustion of nitrate-N beyond 6 hours of incubation, inhibited COD and the phosphate removal rate which was further revived followed by a fresh supply of nitrate-N to the reaction medium. Adequate nitrate-N availability in the system was therefore found to be a prerequisite to support the continuity of denitrification process under a high carbon load. Presence of nitrate-N in the medium did not hamper PHB accumulation. Further nitrate removal rate was observed to follow a first order kinetics. pH of the reaction medium during the course of the experiment supported the prevailing denitrifying conditions in the medium. Initial carbon, nitrate and phosphate concentrations were therefore found to be interdependent parameters that ultimately determined the fate of the reactor system. A simultaneous high COD/NO_3 and low phosphate/COD ratio (in ICL-5) was detrimental for nutrient removal as well as PHB accumulation. NMR and FTIR analysis confirmed that the extracted polymer was PHB. 16S rDNA analysis of the microbial consortia of the seed sludge confirmed the dominance of *Corynebacterium, Rhodocyclus, Paraccocus solventivorance, Paracoccus pantotrophus* and *Paracoccus denitrificans* which are well known for PHB as well as phosphate accumulation using nitrate as an electron acceptor and can provide an economic and sustainable platform for maximum PHB accumulation along with nutrient removal. Rarefaction curve indicated complete bacterial population and corresponding number of OTUs through sequence analysis. Chao1 and Shannon index (H') was used to study the diversity of

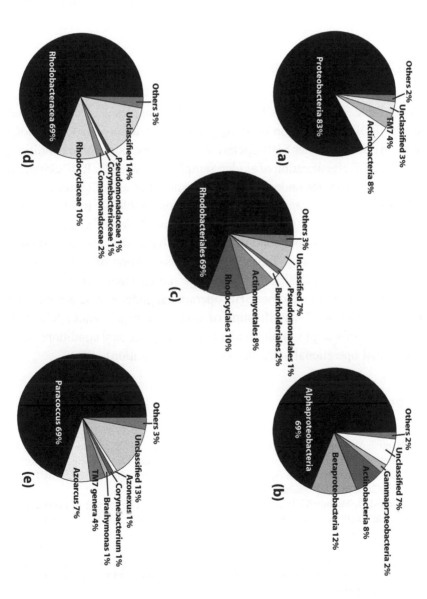

Figure 9: (A) Phylum; (B) Class; (C) Family; (D) Order; (E) Genus.

sampling and the values of Chao1 index supported the results of rarefaction curve. A continuous increase in the unclassified sequences from phylum to family indicates the presence of some novel families or genus of bacteria in the consortium, contributing to the dynamism of the microbial population in the reactor.

5.4 SUPPORTING INFORMATION

S1 File. Table S1. RDP classification of sludge samples used in batch studies. S1a: RDP classification of sludge samples under changed operational conditions (from another SBR operational under anoxic-aerobic condition). Table S2. RDP classification of sludge samples (Class). S2a: RDP classification of sludge samples under changed operational conditions (from another SBR operational under anoxic-aerobic condition). Table S3. RDP classification of sludge samples (Order). S3a: RDP classification of sludge samples under changed operational conditions (from another SBR operational under anoxic-aerobic condition). Table S4. RDP classification of sludge samples (Family). S4a: RDP classification of sludge samples under changed operational conditions (from another SBR operational under anoxic-aerobic condition). Table S5. RDP classification of sludge samples (Genus). S5a: RDP classification of sludge samples under changed operational conditions (from another SBR operational under anoxic-aerobic condition).

REFERENCES

1. Beun J, Verhoef EV, Van Loosdrecht MCM, Heijnen JJ (2000) Stoichiometry and kinetics of Poly- β-Hydroxybutyrate metabolism under Denitrifying conditions in Activated sludge culture. Biotechnol Boieng 68:496–507. doi: 10.1002/(SICI)1097-0290(20000605)68:5%3C496::AID-BIT3%3E3.0.CO;2-S
2. Dionisi D, Majone M, Ramadori R, Becarri M (2001) The storage of acetate under anoxic condition. Water Res 35: 2661–2668. doi: 10.1016/S0043-1354(00)00562-5 PMID: 11456165
3. Kumar SM, Mudliar SN, Reddy KMK, Chakrabarti T (2004) Production of biodegradable plastics from activated sludge generated from a food processing industrial waste water treatment plant. Bioresour Technol 95: 327–330. doi: 10.1016/j.biortech.2004.02.019

4. Tsuneda S, Ohno T, Soejima K, Hirata A (2006) Simultaneous nitrogen and phosphorous removal using denitrifying phosphate accumulating organisms in sequencing batch reactor. Biochem Eng J 27: 191–196. doi: 10.1016/j.bej.2005.07.004

5. Zou H, Du G, Ruan W, Chen J (2006) Role of nitrate in biological phosphorous removal in a sequencing batch reactor. World J Microb Biot 22:701–706. doi: 10.1007/s11274-005-9093-1

6. Bengtsson S, Werker A, Christensson M, Welander T (2008) Production of polyhydroxyalkanoates by activated sludge treating a paper mill wastewater Bioresour Technol 99:509–516.

7. Wallen LL, Rohwedder WK (1974) Poly-β-hydroxyalkanoate from activated sludge. Enviro Sci Technol 8:576–579. doi: 10.1021/es60091a007

8. Seviour RJ, Mino T, Onuki M (2003) The microbiology of biological phosphorous removal in activated sludge systems. FEMS microbiol Rev 27:99–127. Rodgers M,Wu G (2010) Bioresour Technol101:1049–53 doi: 10.1016/S0168-6445(03)00021-4 PMID: 12697344

9. Takabatake H, Satoh H, Mino T, Matsuo T (2000) Recovery of biodegradable plastics from activated sludge process. Water Sci Technol 42:351–356.

10. Kasemsap C, Wantawin C (2007) Batch production of polyhydroxyalkanoate by low-polyphosphate- content activated sludge at varying pH. Bioresour Technol 98: 1020–1027. doi: 10.1016/j.biortech. 2006.04.035 PMID: 16790345

11. Lemos PC, Serafim LS, Reis MAM (2006) Synthesis of polyhydroxyalkanoets from diffret short-chain fatty acids by mixed cultures submitted to aerobic dynamic feeding. J Biotech 122:226–238. doi: 10.1016/j.jbiotec.2005.09.006

12. Perez-Feito R, Noguera DR (2006) Recovery of polyhydroxyalkanoet from activated sludge in an en- hanced biological phosphorous removal bench-scale reactor. Water Environ Res 78:770–775. doi: 10.2175/106143006X99803 PMID: 16929649

13. Schuler AJ, Jenkins D (2003) Enhanced biological phosphorous removal from waste water by biomass with different phosphorous content part-I: Experimental results and comparison with metabolic models Water Environ Res 75:485–495.

14. Zhou Y, Pijuan M, Zeng RJ, Lu H, Yuan Z (2008) Could Polyphosphate Accumulating Organisms (PAOs) be Glycogen-Accumulating Organisms (GAOs)? Water Res 42: 2361–2368. doi: 10.1016/j. watres.2008.01.003 PMID: 18222522

15. Barak Y, Rijn J (2000) A typical polyphosphate accumulation by the denitrifying bacterium Paracocous denitrificance. Appl Env Microbiol 66:1209–1212. doi: 10.1128/AEM.66.3.1209-1212.2000

16. Meinhold J, Arnold E, Isaacs S (1999) Effect of nitrite on anoxic phosphate uptake In biological phos- phorus removal activated Sludge. Water Res 33:1871–1883. doi: 10.1016/S0043-1354(98)00411-4

17. Chua ASM, Takabatake H, Satoh H, Mino T (2003) Production of polyhydoxyalkanoets (PHA) by activated sludge treating municipal waste water:ettect of pH, sludge retention time(SRT), and acetate concentration in influent. Water Res 37:3602–3611. doi: 10.1016/S0043-1354(03)00252-5 PMID: 12867326

18. Salehizadeh H, Van Loosdrecht MCM (2003) Production of polyhydroxyalkanoates by mixed culture: recent trends and biotechnological importance. Biotechnol Adv 22:261–279. doi: 10.1016/j. biotechadv.2003.09.003

19. Kuba T, Wachtmeister A, van Loosdrecht MCM, Heijnen JJ (1994) Effect of nitrate on phosphorus release in biological phosphorus removal systems. Water Sci Technol 30: 263–269.

20. Wentzel MC, Ekama GA, Loewenthal RE, Dold PL, Marais GvR (1989) Enhanced polyphosphate organism culture in activated sludge system. Part II. Experimental behaviour. Water SA 15:71–88.

21. Smolders GJF, van der Meij J, van Loosdrecht MCM, Heijnen JJ (1994) Stoichiometric model of the biological phosphorous removal process. Biotechnol Bioeng 44: 837–848. doi: 10.1002/bit.260440709 PMID: 18618851

22. APHA (1998) Standard methods for the examination of water and waste water. 20th Ed. American public health association/water environment federation. Washington, D.C.USA.

23. Valappil SP, Misra SK, Boccaccini AR, Keshavarz T, Bucke C, et al. (2007) Large-scale production and efficient recovery of PHB with desirable material properties, from the newly characterised Bacillus Cereus SPV. J Biotechnol 132: 251–258. doi: 10.1016/j.jbiotec.2007.03.013 PMID: 17532079

24. Cole JR, Wang Q, Cardenas E, Fish J, Chai B, et al. (2009) The Ribosomal Database Project: improved alignments and new tools for rRNA analysis. Nucleic Acids Res 37 (Database issue): D141–145. doi: 10.1093/nar/gkn879 PMID: 19004872

25. Wright ES, Yilmaz LS, Noguera DR (2011) DECIPHER- A Search-Based Approach to Chimera Identifi- cation for 16S rRNA Sequences. Appl Environ Microb 78: 717–725. doi: 10.1128/AEM.06516-11

26. Nawrocki EP, Eddy SR (2007) Query-Dependent Banding (QDB) for Faster RNA Similarity Searches. PLoS Comput Biol 3:540–554. doi: 10.1371/journal.pcbi.0030056

27. Fernandez-Nava Y, Maranon E, Soons J, Castrillon L (2010) Denitrification of high nitrate-N concentration wastewater using alternative carbon source. J Hazard Mater 173: 682–688. doi: 10.1016/j. jhazmat.2009.08.140 PMID: 19782470

28. Qingjuan M, Fenglin Y, Lifen L, Fangang M (2008) Effect of COD/N ratio and DO concentration on simultaneous nitrification and denitrification in an airlift internal circulation membrane bioreactor. J Environ Sci 20:933–939. doi: 10.1016/S1001-0742(08)62189-0

29. Fogler L, Briski F, Sipos L, Vucovic M (2004) High nitrate removal from synthetic waste water with the mixed bacterial culture. Bioresource technol:1–10.

30. Rene ER, Kim SJ, Park HS (2007) Effect of COD/N ratio and salinity on the performance of sequencing batch reactor. Biresour Technol 4:839–846.

31. Zeng RJ, Yuan Z, Keller J (2003) Enrichment of denitrifying glycogen accumulating organisms in anaerobic/anoxic activated sludge system. Biotechnol Bioeng 81:398–404.

32. Crocetti GR, Banfield JF, Keller J, Bond PL, Black LL (2002). Glycogen accumulating organism in laboratory sacle waste water treatment process. Microbiology 148, 3353–3364. PMID: 12427927

33. Shi J, Lu X, Yu R, Gu Q, Zho Y (2014) Influence of wastewater composition on nutrient removal behaviors in the new anaerobic–anoxic/nitrifying/induced crystallizationprocess. Saudi J Biol Sci 21:71–80. doi: 10.1016/j.sjbs.2013.06.003 PMID: 24596502

34. Zhang SH, Huang Y, Hua YM (2010) Denitrifying dephosphatation over nitrite: effects of nitrite concentration, organic carbon, and pH. Bioresour Technol 101: 3870–3875. doi: 10.1016/j.biortech.2009.12. 134 PMID: 20110164

35. Hu ZR, Wentzel MC, Ekama GA (2002) Anoxic growth of phosphate accumulating organism (PAOs) in biological nutrient removal activated sludge system. Water Res 36: 4927–4937. doi: 10.1016/S0043- 1354(02)00186-0 PMID: 12448537

36. Harper WF, Jenkins D (2003) The effect of an initial anaerobic zone on the nutrient requirements of activated sludge. Water Env Res 75:216–224. doi: 10.2175/106143003X140999

37. Bassin JP, Kleerebezem R, Dezotti M, van Loosdrecht MCM (2012) Simultaneous nitrogen and phosphorous removal in aerobic granular sludge reactor operated at different temperature. Water Res 46:3805–3816. doi: 10.1016/j.watres.2012.04.015 PMID: 22591819

38. Freitas F, Temudo MF, Carvelhi G, Oheman A, Reis MAM (2009) Robustness of sludge enriched with short SBR cycle foe biological nutrient removal. Bioresour technol 6: 1969–1976. doi: 10.1016/j. biortech.2008.10.031

39. Jena J, Ray S, Pandey S, Das T (2013) Effect of COD/N ratio on simultaneous removal of nutrients and COD from synthetic high strength waste water under anoxic conditions. J Sci Ind Res 72: 127–131.

40. Merzouki M, Bernet N, Delgenes JP, Moletta R, Benlemlih M (2001) Effects of operating parameters on anoxic biological phosphorous removal in anaerobic anoxic sequencing batch reactor. Environ Technol 22:397–408. doi: 10.1080/09593332208618269 PMID: 11329803

41. Mulkerrins D, Dobson ADW, Colleran E (2004) Parameters affecting biological phosphate removal from wastewaters. Environ Int 30:249–259. doi: 10.1016/S0160-4120(03)00177-6 PMID: 14749113

42. Hu JY, Ong SL, Ng WJ, Lu F, Fan XJ (2003) A new method for characterizing denitrifying phosphorous removal bacteria by using three different types of electron acceptor. Water Res 37: 3463–3471. doi: 10.1016/S0043-1354(03)00205-7 PMID: 12834739

43. Jhonson K, Kleerebezem R, van Loosdrecht MCM (2010) Influence of C/N ratio on the performance of polyhydroxybutyrate (PHB) producing sequencing batch reactor. Water. Res 44:2141–2152. doi: 10. 1016/j.watres.2009.12.031

44. Arun V, Mino T, Matsuo T (1989) Metabolism of carboxylic acids located in and around the glycolytic pathway and the TCA cycle in the biological phosphorous removal process. Water Sci Technol 21: 363–374.

45. Watanabe T, Motoyama H, Kuroda M (2001) Denitrification and neutralization treatment by direct feeding of an acidic wastewater containing copper ion and high-strength nitrate to a bio-electrochemical reactor process. Water res 17:4102–10. doi: 10.1016/S0043-1354(01)00158-0

46. Reddy MV, Mohan SV (2012) Influence of aerobic and anoxic microenvironments on polyhydroxyalkanoates (PHA) production from food waste and acidogenic effluents using aerobic consortia. Bioresour Technol 103: 313–321. doi: 10.1016/j. biortech.2011.09.040 PMID: 22055090

47. Ahn J, Daidou T, Tusneda S, Hirata A (2002) Transformation of phosphorous and relevant intracellular compounds by a phosphorous-accumulating enrichment culture in the presence of both electron acceptor and electron donor. Biotechnol Bioeng 79: 83–93. doi: 10.1002/bit.10292 PMID: 17590934

48. Wang Y, Geng J, Ren Z, Guo G, Wang C, et al. (2012) Effect of COD/N and COD/P on the PHA transformation and dynamics of microbial community structure in a denitrifying phosphorous removal process. J Chem Technol Biotecnol 88:1228–1236. doi: 10.1002/jctb.3962

49. Serafim LS, Lemos PC, Albuquerque MGE, Reis MAM (2008) Strategies for PHA production by mixed cultures and renewable waste materials. Appl Microbiol Biot 81: 615–628. doi: 10.1007/s00253-008- 1757-y

50. Bucci V, Majed N, Hellweger FL, Gu AZ (2012) Heterogeneity of Intracellular Polymer Storage States in Enhanced Biological Phosphorus Removal (EBPR)—observation and Modeling. Environ Sci Technol 46: 3244–3252. doi: 10.1021/es204052p PMID: 22360302

51. Kathiraser Y, Aroua MK, Ramchandran KB, Tan IKP (2007) Chemical characterization of medium- chain-length polyhydroxyalkanoets (PHA) recovered by enzymatic treatment and ultrafiltration. J Chem Technol Biot 82: 847–855. doi: 10.1002/jctb.1751

52. Samantaray S, Nayak JK, Mallick N (2011) Waste water utilization for Poly- ß-Hydroxylbutyrate production by cyanobacterium *Aulosira Fertilissima* in a recirculatory aquaculture system. Appl Environ Microb 77: 8735–8743. doi: 10.1128/AEM.05275-11

53. Valappil SP, Misra SK, Boccaccini AR, Keshavarz T, Bucke C, et al. (2007) Large-scale production and efficient recovery of PHB with desirable material properties, from the newly characterised Bacillus Cereus SPV. J Biotechnol 132: 251–258. doi: 10.1016/j.jbiotec.2007.03.013 PMID: 17532079

54. Liao R, Shen K, Li AM, Shi P, Li Y, et al. (2013) High-nitrate wastewater treatment in an expanded granular sludge bed reactor and microbial diversity using 454 pyrosequencing analysis. Bioresour Technol 134:190–197. doi: 10.1016/j.biortech.2012.12.057 PMID: 23500551

55. Kong Y, Xia X, Nielsen JL, Nielsen PH (2007) Structure and function of the microbial community in a full-scale enhanced biological phosphorus removal plant. Microbiology 153: 124061–4073. doi: 10.1099/mic.0.2007/007245-0

56. Hu M, Wang X, Wen X, Xia Y (2012) Microbial community structure in different waste water treatment plants as revealed by 454-pyrosequencing analysis. Bioresour Technol 117:72–79. doi: 10.1016/j. biortech.2012.04.061 PMID: 22609716

57. Juretschkoa S, Loy A, Lehner A, Wagner M (2002) The Microbial Community Composition of a Nitrifying-Denitrifying Activated Sludge from an Industrial Sewage Treatment Plant Analyzed by the Full-Cycle rRNA Approach. Systm. Applied Microbial 1:84–99. doi: 10.1078/0723-2020-00093

58. Wagner M, Loy A (2002) Bacterial community composition and function in sewage treatment system. Curr Opin Biotechnol 13: 218–227. doi: 10.1016/S0958-1669(02)00315-4 PMID: 12180096

59. Zhang T, Shao M, Ye L (2012) 454 Pyrosequencing reveals bacterial diversity of activated sludge from 14 sewage treatment plants. ISME J 6:1137–1147. doi: 10.1038/ismej.2011.188 PMID: 22170428

60. Kragelund C, Kong Y, Waarde J, Thelen K, Eikelboom D, et al. (2006) Ecophysiology of different filamentous *Alphaproteobacteria* in industrial wastewater treatment plants. Microbiology 152: 3003–3012. doi: 10.1099/mic.0.29249-0 PMID: 17005981

61. Ferguson ST, Gadian DG (1979) Evidence from 31P magnetic resonance that polyphosphate synthesis is a slip reaction in *Paracoccus denitrificance*. Biochem Soc Trans 7:176:179. PMID: 437269

62. Kuba T, Smolders G, van Loosdrecht MCM, Heijnen JJ (1993) Biological phosphorous removal from waste water by anaerobic-anoxic sequencing batch reactor. Water Sci Technol 27:241–245.

63. Mino T, Liu WT, Kurisu F, Matsuo T (1995) Modelling glycogen storage and denitrification capability of microorganism in enhanced biological phosphorous removal process. Water Sci Technol 31:25–34. doi: 10.1016/0273-1223(95)00177-O

64. Shi HP, Lee CM (2006) Combining anoxic-denitrifying ability with aerobic-anoxic phosphorus removal examination to screen identifying phosphorous removing bacteria. Int Biodeter Biodegr 57:121–128. doi: 10.1016/j.ibiod.2006.01.001

65. Nielsen PH, Mielczarek AT, Kragelund C, Nielsen JL, Saunders AM, et al. (2010). A conceptual ecosystem model of microbial communities in enhanced biological phosphorus removal plants. Water Res 44:5070–5088. doi: 10.1016/j.watres.2010.07.036 PMID: 20723961

66. Gebremariam SY, Beutel MW, Christian D, Hess TF (2011) Research advances and challenges in the microbiology of enhanced biological phosphorus removal-A critical review. Water Environ Res 83:195–219. doi: 10.2175/106143010X1278028862 8534 PMID: 21466069

67. Kapagiannidis AG, Zafiriadis I, Aivasids A (2012) Effect of basic operating parameters on biological phosphorus removal in a continuous-flow anaerobic-anoxic activated sludge system. Bioprocess Biosyst Eng 35:371–382. doi: 10.1007/s00449-011-0575-2 PMID: 21796365

68. Ye L, Zhang T (2011) Pathogenic bacteria in sewage treatment plants as revealed by 454 Pyrosequencing. Environ Sci Technol 45:7173–7179. doi: 10.1021/es201045e PMID: 21780772

69. Zhou S, Catherine C, Rathnasingh C, Somasundar A, Park S (2013) Production of 3-Hydroxypropionic Acid from Glycerol by Recombinant Pseudomonas Denitrificans. Biotechnol Bioeng 110:3177–3187. doi: 10.1002/bit.24980 PMID: 23775313

70. Van Loosdrecht MCM, Pot MA, Heijnen JJ (1997) Importance of bacterial storage polymers in bioprocess. Water Sci Technol 35:41–47. doi: 10.1016/S0273-1223(96)00877-3

There are several supplemental files that are not available in this version of the article. To view this additional information, please use the citation on the first page of this chapter.

CHAPTER 6

Characterization of Pure Cultures Isolated from Sulfamethoxazole-Acclimated Activated Sludge with Respect to Taxonomic Identification and Sulfamethoxazole Biodegradation Potential

BASTIAN HERZOG, HILDE LEMMER, HARALD HORN
AND ELISABETH MÜLLER

6.1 BACKGROUND

The widespread usage, disposal all around the world and a consumption of up to 200,000 t per year, makes the various groups of antibiotics an important issue for micropollutants risk assessment [1,2]. Their discharge and thus presence in the environment has become of major concern for environmental protection strategies. Antibiotics are designed to inhibit microorganisms and therefore influence microbial communities in different ecosystems [3,4]. Monitoring programs have already shown that antibiotics can be found nearly everywhere in the environment, even in concentrations

Characterization of Pure Cultures ilolated from Sulfamethoxazole-Acclimated Activated Sludge with Respect to Taxonomic Identification and Sulfamethoxazole Biodegradation Potential. © Herzog B, Lemmer H, Horn H, and Müller E, licensee BioMed Central Ltd. BMC Microbiology **13** (2013), doi:10.1186/1471-2180-13-276. Licensed under a Creative Commons Attribution License, http://creativecommons.org/licenses/by/2.0/.

up to μg L^{-1} leading to antibiotic resistance in organisms [5-9]. Antibiotic resistance genes might be transferred to human-pathogenic organisms by horizontal gene-transfer and become a serious issue, especially multidrug resistance in bacteria [10-12].

Sulfamethoxazole (SMX) is one of the most often applied antibiotics [13]. The frequent use of SMX results in wastewater concentrations up to μg L^{-1} and surface water concentrations in the ng L^{-1} scale [14-17]. Even in groundwater SMX was found at concentrations up to 410 ng L^{-1} [16]. These SMX concentrations might be too low for inhibitory effects as the MIC$_{90}$ for *M. tuberculosis* was found to be 9.5 mg L^{-1} [18], but they might be high enough to function as signalling molecule to trigger other processes like quorum sensing in environmental microbial communities [19].

As shown by different studies [20-23], SMX can induce microbial resistances and reduce microbial activity and diversity arising the need for a better understanding of SMX biodegradation. SMX inflow concentrations in WWTPs in μg L^{-1} combined with often partly elimination ranging from 0% to 90% [4,6,15,24] result in high effluent discharge into the environment. To predict the extent of removal knowledge about the responsible biodegrading microorganisms is implicitly required to optimize environmental nutrient conditions for SMX removal and degradation rates. It is known that SMX can be removed by photodegradation occurring mainly in surface waters [25,26] and sorption processes in activated sludge systems [27]. However, biodegradation is, especially in WWTPs, probably the major removal process. Literature data focusing on SMX biodegradation in lab scale experiments with activated sludge communities and pure cultures showed a high fluctuation from almost complete SMX elimination [9, 28, 29] to hardly any removal of SMX [30]. The determined SMX biodegradation potential was clearly affected by nutrient supply. Therefore this study's emphasis is on clarifying the effect that addition of readily degradable carbon and/or nitrogen sources in some cases significantly enhanced SMX elimination [31] while in other cases supplementation showed no effect [28].

For this purpose pure culture were isolated from SMX-acclimated activated sludge communities and identified in respect to taxonomy and biodegradation capacity. Aerobic SMX biodegradation experiments with different species were carried out at various nutrient conditions to screen biodegradation potential and behaviour as a base for future research on biodegradation pathways.

6.2 RESULTS

6.2.1 SMX BIODEGRADATION

6.2.1.1 CULTIVATION AND EVALUATION OF PURE CULTURES BIODEGRADATION POTENTIAL

Isolation of pure cultures was accomplished from SMX-acclimated ASC. Growth of cultures on solid R2A-UV media, spiked with 10 mg L^{-1} SMX, was controlled every 24 hours. All morphologically different colonies were streaked onto fresh R2A-UV agar plates, finally resulting in 110 pure cultures. For identification of potential SMX biodegrading cultures, all 110 isolates were inoculated in 20 mL MSM-CN media. SMX biodegradation was controlled every two days. After two days a decrease in absorbance was already detected in 5 cultures followed by 7 more at day 4 and 6 while the remaining cultures showed no change. The experiment was stopped after 21 days revealing no further SMX biodegrading culture. A 50% cutoff line defined a 50% decrease in UV-absorbance being significant enough to be sure that the corresponding organisms showed biodegradation. 12 organisms showed a decrease in absorbance greater than 50% of initial value and were defined as potential SMX biodegrading organisms. They were taxonomically identified and used for subsequent biodegradation experiments.

Additionally, biodegradation of these 12 identified isolates was validated by LC-UV (Table 1). For cost efficiency only initial and end concentrations of SMX in the media were determined as absorbance values did not change any more. A decrease in SMX concentration from initially 10 mg L^{-1} to below 5 mg L^{-1} was detected for all 12 isolates (Table 1) after 10 days of incubation. It was proven that only 3 cultures eliminated all 10 mg L^{-1} SMX completely while the residual SMX concentrations for the remaining cultures ranged from 0.23 to 4.35 mg L^{-1} after 10 days of incubation.

6.2.1.2 TAXONOMIC AND PHYLOGENETIC IDENTIFICATION OF PURE CULTURES

All 12 cultures were identified by 16S rRNA gene sequence analysis to evaluate their phylogenetic position and closest relative. Four cultures, SMX 332, 333, 336 and 344, turned out to be the same organism closely related to *Pseudomonas* sp. He (AY663434) with a sequence similarity of 99%. Only SMX 344 was kept for further experiments as it showed fastest biodegradation in pre-tests (Table 1). Hence, a total of 9 different bacterial species with SMX biodegradation capacity were obtained. Their accession numbers, genus names and their closest relatives as found in the NCBI database (http://blast.ncbi.nlm.nih.gov/Blast.cgi), are shown as a maximum likelihood-based phylogenetic tree (Figure 1) evaluated with 16S rRNA gene sequence comparisons to calculate the most exact branching [28].

Table 1: Initial and end concentrations of SMX accomplished with 12 biodegrading pure culture isolates that were gained out of 110 cultures

Pure culture	SMX conc. after 10 days [mg L^{-1}]
Brevundimonas sp. SMXB12	0.00
Microbacterium sp. SMXB24	0.00
Microbacterium sp. SMX348	0.00
Pseudomonas sp. SMX321	0.68
Pseudomonas sp. SMX330	0.68
Pseudomonas sp. SMX331	2.68
Pseudomonas sp. SMX 333*	1.09
Pseudomonas sp. SMX 336*	4.35
Pseudomonas sp. SMX 342*	1.09
Pseudomonas sp. SMX344*	0.23
Pseudomonas sp. SMX345	1.58
Variovorax sp. SMX332	3.53

*duplicate organisms. All but SMX344 were discarded. Taxonomic identification succeeded with BLAST (http://blast.ncbi.nlm.nih.gov/Blast.cgi).

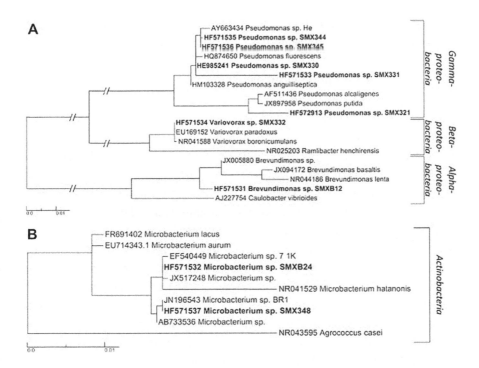

Figure 1: Maximum likelihood-based trees reflecting the phylogeny and diversity of the isolated nine species capable of SMX biodegradation based on nearly complete 16S rRNA gene sequence comparisons. Phylogenetic tree calculated for A) *Pseudomonas* spp., *Variovorax* spp. and *Brevundimonas* spp. and B) for *Microbacterium* spp.. The tree shows the sequences obtained in this study (bold text) and their next published relatives according to the NCBI database (plain text). Numbers preceding taxonomic names represent EMBL sequence accession numbers. Scale bar indicates 0.01% estimated sequence divergence.

Seven of the nine isolates are affiliated within the phylum *Proteobacteria* represented by the classes *Alpha-*, *Beta-* and *Gammaproteobacteria*, while two belonged to the Phylum *Actinobacteria*.

The phylogenetic positions of the seven isolated pure cultures, affiliated within the phylum *Proteobacteria*, were located in the same tree (Figure 1A). Five different *Pseudomonas* spp. were identified and form two different clades representing a highly diverse group. *Pseudomonas* sp. SMX344 and 345 is building an individual cluster but belonged to the same group as

SMX330 and 331. All four are closely related to *P. fluorescens* but SMX331 showed a remarkable difference. In contrast to the described *Pseudomonas* spp. above, *Pseudomonas* sp. SMX321 clusters together with *P. putida* and *P. alcaligenes* but forms an individual branch.

The other two *Proteobacteria* identified pure cultures belonged to the genera *Variovorax* (SMX332) and *Brevundimonas* (SMXB12). The isolated *Variovorax* SMX332 fell into the *Variovorax paradoxus/boronicumulans* group with a sequence similarity >99% to *V. paradoxus* (EU169152).

The *Brevundimonas* sp. SMXB12 was clearly separated from its closest relatives *Brevundimonas basaltis* and *B. lenta* and formed its own branch.

Both *Actinobacteria* affiliated pure cultures were identified as *Microbacterium* spp. and were embedded in a new phylogenetic tree as their phylogenetic position was too far from the other isolates (Figure 1B). The two isolated species were affiliated to two different clades clearly separated from *M. lacus* and *M. aurum. Microbacterium* sp. SMXB24 fell into the same group as *Microbacterium* sp. 7 1 K and *M. hatatonis* but the branch length clearly showed separation. *Microbacterium* sp. SMX348 was closely related with a sequence similarity of >99% to *Microbacterium* sp. BR1 which was found to biodegrade SMX in an acclimated membrane bioreactor [29].

6.2.2 SMX BIODEGRADATION STUDIES WITH PURE CULTURES

Setups with sterile biomass (heat-killed) and without biomass (abiotic control) proved SMX to be stable under the operating conditions. Therefore sorption onto biomass or other materials was shown to be negligible. Photodegradation was excluded by performing all experiments in the dark.

To characterize biodegradation ability and rate and evaluate an optimal nutrient environment for SMX utilization of the isolated and identified 9 pure cultures, subsequent experiments were performed. In the presence of readily degradable carbon and/or nitrogen sources (Figures 2 and 3) SMX was faster biodegraded compared to setups with SMX as sole carbon/nitrogen source (Figure 3). 54 setups (three media for each of the 9 cultures in duplicate setups) with different nutrient compositions were set up and SMX biodegradation rates were evaluated using UV-AM values (Table 2). Different SMX biodegradation patterns were observed proving that the presence or absence of readily degradable and complex nutrients significantly influenced biodegradation.

R2A-UV media were sampled once a day as it was assumed that biodegradation might be faster compared to the other two nutrient-poor media. Biodegradation rates of 2.5 mg L^{-1} d^{-1} were found for all nine species not showing any different biodegradation behaviors or patterns (Figure 4A). Although biomass growth affected background absorbance that increased with cell density, UV-AM could still be applied to monitor biodegradation as background absorbance was still in a measurable range.

In MSM-CN (Figure 2), offering only specific C- and N-sources, the biodegradation rates ranged from 1.25 to 2.5 mg L^{-1} d^{-1} (deviations between the duplicate setups were below 1%) showing clear differences for the different species, even for the five *Pseudomonas* spp.. While *Pseudomonas* sp. SMX321 biodegraded SMX with 2.5 mg L^{-1} d^{-1}, *Pseudomonas* sp. SMX344 just showed a rate of 1.25 mg L^{-1} d^{-1}. The same effect was found for the two *Microbacterium* spp.. While *Microbacterium* sp. SMXB12 removed SMX with 1.7 mg L^{-1} d^{-1}, *Microbacterium* sp. SMX348 showed a removal of 1.25 mg L^{-1} d^{-1} only. Biodegradation pattern in MSM-CN of four isolates (SMX321, 345, 348 and B12) revealed a short lag phase of two days with no SMX removal (Figure 2A) while the other five were able to biodegrade SMX already after two days and showed a constant SMX removal during cultivation (Figure 2B).

In MSM (Figure 3), with SMX as sole C- and N- source, the removal rate of SMX was even lower. Biodegradation rates of 1.0 mg L^{-1} d^{-1} were found for *Brevundimonas* sp. SMXB12 while *Pseudomonas* sp. SMX321 showed 1.7 mg L^{-1} d^{-1}. All other species showed removal rates of 1.25 mg L^{-1} d^{-1}. These experiments with SMX as sole C/N-source proved that it could serve as nutrient source but with up to 2.5-fold reduced biodegradation rates. Biodegradation pattern in MSM was similar to that in MSM-CN with a lag phase of two days for the four isolates SMX321, 345, 348 and B12 (Figure 3A) and no lag phase for the isolates SMX 330, 331, 332, 344, and B24 starting to utilize SMX already after two days (Figure 3B). In general it was found that the five *Pseudomonas* spp. and the two *Microbacterium* spp. did not show the same biodegradation behavior. At

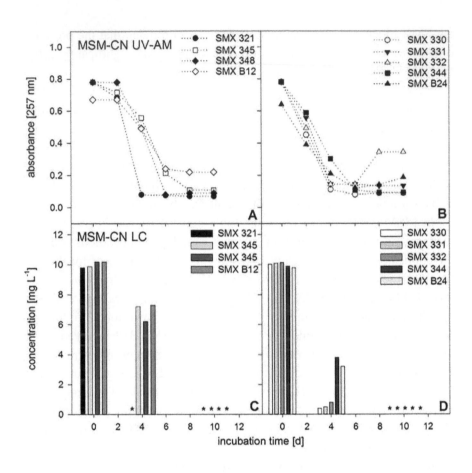

Figure 2: Aerobic SMX biodegradation patterns of pure cultures in MSM-CN media. A, B) measured with UV-AM, initial SMX concentration 10 mg L^{-1}. C, D) LC-UV analyses of SMX concentrations in the used pure cultures in MSM-CN. Determination was performed at experimental startup, after 4 and 10 days to verify UV-AM values. Asterisks indicate measured values below limit of detection. Shown are mean values of SMX absorbance in duplicate experiments. Standard deviations were too low to be shown (<1%).

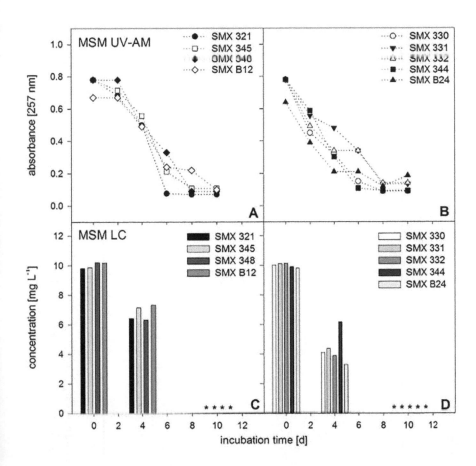

Figure 3: Aerobic SMX biodegradation patterns of pure cultures in MSM media. A, B) measured with UV-AM, initial SMX concentration 10 mg L^{-1}. C, D) LC-UV analyses of SMX concentrations in the pure cultures in MSM at experimental startup, after 4 and 10 days to validate UV-AM. Asterisks indicate measured values below limit of detection. Shown are mean values of SMX absorbance in duplicate experiments. Standard deviations were too low to be shown (<1%).

Table 2: Biodegradation rates of the cultures able to biodegrade SMX

Acession/isolate	Phylum	Biodegradation rate* $(mgL^{-1} \cdot d^{-1})$		
		R2A-UV	MSM-CN	MSM
HF571531, *Brevundimonas* sp. SMXB12	Proteobacteria	2.5	1.7	1.0
HF571532, *Microbacterium* sp. SMXB24	Actinobacteria	2.5	1.25	1.25
HF571537, *Microbacterium* sp. SMX348	Actinobacteria	2.5	1.7	1.25
HF572913, *Pseudomonas* sp. SMX321	Proteobacteria	2.5	2.5	1.7
HE985241, *Pseudomonas* sp. SMX330	Proteobacteria	2.5	1.7	1.25
HF571533, *Pseudomonas* sp. SMX331	Proteobacteria	2.5	1.7	1.25
HF571535, *Pseudomonas* sp. SMX344	Proteobacteria	2.5	1.7	1.25
HF571536, *Pseudomonas* sp. SMX345	Proteobacteria	2.5	1.25	1.25
HF571534, *Variovorax* sp. SMX332	Proteobacteria	2.5	1.7	1.25

*calculated from duplicate experiments (n=2). Standard deviations between duplicate setups were below 1% and are not shown. Isolation was performed from an SMX-acclimated AS community, followed by identification with 16S rRNA sequencing. ENA accession numbers and species names are provided.

least one member of each group always showed a lag phase while the other immediately started SMX biodegradation.

As UV-AM revealed sufficient to monitor SMX biodegradation (Table 1) LC-UV measurements were only performed at the start of the experiment, day 4 and at day 10 as control measurement (Figures 3B, 4C, D). LC-UV showed that in R2A-UV all cultures removed 10 mg L^{-1} SMX in 4 days (Figure 2B) while in MSM-CN only *Pseudomonas* sp. SMX321 removed all SMX within 4 days (Figure 3C). The remaining 8 cultures still showed residual SMX concentrations from 0.4 to 7.3 mg L^{-1} and complete SMX elimination was achieved only at day 10 (Figure 3C, D). In MSM after 4 days SMX was still present in all nine cultures in concentrations above 3.6 mg L^{-1} and only after 10 days SMX was below the limit of detection (Figure 4C, D). LC-UV values could be compared to UV-AM values and proved this simple approach to be applicable for screening SMX biodegradation.

6.3 DISCUSSION AND CONCLUSIONS

This study focused on the cultivation of pure culture SMX biodegrading organisms to perform specific biodegradation experiments. It is known that cultivation, especially on solid media, is affected with the problem described as "viable but non cultivable" (VBNC) [30,31]. Solid media being implicitly required for the isolation of pure cultures is for sure limited in its cultivation efficiency mainly due to reduced water content and different or inappropriate nutrient conditions. Thus only a low percentage of around 1% of the active organisms in environmental samples [32] and around 15% from activated sludge can be cultivated [33,34]. In this study 9 different isolates out of 110 pure cultures were obtained that showed SMX biodegradation. This quite high percentage of almost 10% was only possible with a two-step SMX-acclimation experiment that was conducted to increase the chance to cultivate SMX biodegrading organisms by applying a strong selective pressure using 10 mg L^{-1} SMX in the media. Furthermore, R2A medium that is known to work well for isolation of aquatic organisms [35] was applied for the cultivation of bacteria being assumed to be at least SMX-resistant when growth was observed on SMX-reinforced R2A. However, a lot more organisms compared to those cultivated in this study might be present in activated

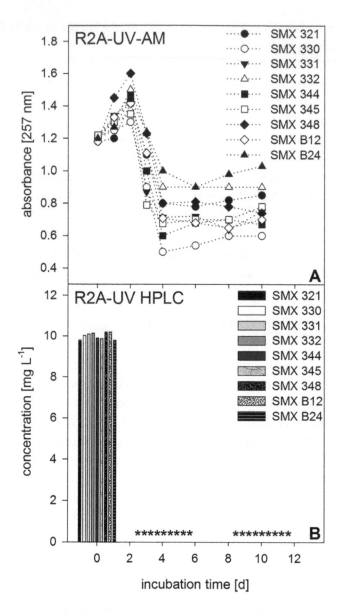

Figure 4: Aerobic SMX biodegradation patterns of pure cultures in R2A-UV media. A) measured with UV-AM, initial SMX concentration 10 mg L-1. B) LC-UV analyses of SMX concentrations within the nine pure cultures in R2A-UV media performed at experimental startup, after 4 and 10 days to verify the results of UV-AM. Asterisks indicate measured values below limit of detection. Shown are mean SMX absorbance values of duplicate experiments. Standard deviations were too low to be shown (<1%).

sludge capable of SMX biodegradation. These VBNCs might be taxonomically characterized by culture-independent methods, e.g. restriction fragment length polymorphism screening [36,37]. However, for our focus on linking biodegradation patterns, rates and nutrient utilization to specific species these methods were not feasible. Only with actively biodegrading pure cultures a clear and precise coherence between SMX biodegradation and taxonomically identified species is possible. As a final goal, pure cultures would allow to analyze species-specific biodegradation products and thus determine potential SMX biodegradation pathways. Applying that knowledge to WWTP techniques would provide a strategy to selectively enhance biodegrading species in activated sludge systems improving and stabilizing SMX removal efficiency.

Therefore phylogenetic identification of potential SMX biodegrading species is implicitly required. As shown in this study five of the nine SMX biodegrading species found belonged to the genus *Pseudomonas* confirming this group to play an important role for the biodegradation of micropollutants. This was proved for e.g. acetaminophen or chlorinated compounds by many other studies [38-40]. Additionally, two isolates SMXB24 and SMX348 were identified as *Microbacterium* sp.. It was shown that *Microbacterium* sp. SMXB24 is closely related to *Microbacterium* sp. 7 1 K, an organism that was found to be related with phytoremediation. The second *Microbacterium* sp. SMX348 is closely related to *Microbacterium* sp. BR1 which was isolated from an acclimated SMX biodegrading membrane bioreactor, proving this species' crucial role for the biodegradation of SMX [29]. In addition the general potential of different *Microbacteria* species for the biodegradation of xenobiotic compounds has been highlighted in the literature [41,42]. Also *Variovorax paradoxus*, closely related to the isolated *Variovorax* sp. SMX332, is known from literature to be capable of biodegrading a large variety of pollutants including sulfolene and other heterocyclic compounds [43]. Therefore it seems likely that the isolated *Variovorax* sp. SMX332 might also be able to biodegrade SMX. Finally, also for the group *Brevundimonas* spp. some literature data exist proving that these organisms might play a role in the removal of antibiotics [44].

Taxonomic identification was followed by observing influences on biodegradation rate and efficiency due to the availability of nutrients. Biodegradation rates decreased with reduced nutrient content from the complex

R2A-UV over nutrient-poor MSM-CN and MSM media and more time was needed to remove SMX. MSM media contained SMX as sole carbon and nitrogen source at a concentration of 10 mg L^{-1} and thus provided just around 4.8 mg L^{-1} carbon and 1.7 mg L^{-1} nitrogen. These conditions, with SMX being the only nutrient in MSM, showed an effect on biodegradation and reduced removal efficiency but proved the organisms' ability to utilize SMX as sole nutrient and/or energy source. However, this indicates that complex nutrients and higher nutrient concentrations seem to have a positive effect on biodegradation due to cometabolic [45] or diauxic effects [46] as the very high SMX removal rates of 2.5 mg L^{-1} d^{-1} confirmed that they were significantly higher than the one of 0.0079 mg L^{-1} d^{-1} found in a previous study [47].

In general, SMX biodegradation might be based more on a diauxic process, i.e. readily degradable nutrients are used up first followed by SMX utilization, rather than real co-metabolism, i.e. two substrates are used up in parallel when provided together, as experiments with R2A-UV media showed. A strong increase in UV-AM, attributed to biomass growth due to a fast nutrient consumption pro- vided by the complex R2A-UV media, was followed by a rapid SMX elimination. In MSM-CN or MSM, as the nutrients concentrations were too low to foster excessive biomass growth, such an increase was not observed. Even at low cell densities SMX was rapidly removed proving that biomass concentration is not as important as cellular activity. Therefore, the higher removal rates in presence of sufficient nutrients also showed that SMX biodegradation was a rapid and complex metabolic process.

Therefore, information about the biodegradation potential of the isolated bacterial strains with respect to the availability of nutrients might increase the elimination efficiency in WWTPs as the treatment process could be specifically adapted to the needs of the biodegrading species.

For future research, the availability of isolated species will allow screening for biodegradation intermediates and/or stable metabolites and determination of species-specific biodegradation pathways. To date only few data on SMX metabolites such as 3-amino-5-methyl-isoxazole found in SMX degrading activated sludge communities [48] and hydroxy-N-(5-methyl-1,2-oxazol-3-yl)benzene-1- sulfonamide detected in an SMX degrading consortium of fungi and *Rhodococcus rhodochrous* exists [45]. Further research is also needed to screen for the nutrient influence on metabolite formation, i.e. if the isolated pure cultures produce different metabolites due to changing nutrient conditions.

6.4 METHODS

6.4.1 CHEMICALS AND GLASSWARE

Sulfamethoxazole (SMX, 99.8% purity) was purchased from Sigma Aldrich (Steinheim, Germany), all other organic media components were from Merck KGaA (Darmstadt, Germany) while the inorganic media components were purchased from VWR (Darmstadt, Germany). High-purity water was prepared by a Milli-Q system (Millipore, Billerica, MA, USA). All glassware used was procured from Schott AG (Mainz, Germany) and pre-cleaned by an alkaline detergent (neodisher®, VWR Darmstadt, Germany) followed by autoclaving for 20 min at 121°C.

6.4.2 ACTIVATED SLUDGE SAMPLING

Activated sludge (AS) was taken as grab sample from stage 1 of a 2-stage municipal conventional activated sludge plant (CAS-M), located near the city of Munich, Germany and treating 1 million populations equivalents. Stage 1 is the high load stage with a food to microorganism ratio of 0.64 kg BOD_5 kg^{-1} $MLSS^{-1}$. The influent consists of municipal and industrial wastewater (1:1). 500 mL AS (SMX concentration 600 ng L^{-1}) were collected in

pre-cleaned 1 L glass bottles, stored at 4°C and used within 24 h for inoculation of the different setups.

6.4.3 EXPERIMENTAL SETUP

6.4.3.1 SMX ACCLIMATED ASC

Evaluation of AS biodegradation potential obtained from the WWTP, was
performed in 150 mL R2A-UV media (casein peptone 1,000 mg L^{-1}, glucose
500 mg L^{-1}, potassium phosphate 300 mg L^{-1}, soluble starch 300 mg L^{-1},
DOC:N ratio 7:1, pH 7.4), spiked with 10 mg L^{-1} SMX to apply a high selective pressure. Non-SMX-resistant organisms were ruled out and the chance
to obtain SMX biodegrading organisms was increased in subsequent isolation steps. After biodegradation occurred the experiment was stopped and
the remaining biomass was used to inoculate a second setup under the same
conditions to further decrease microbial diversity and favor SMX-resistant/
biodegrading organisms. After the second setup showed biodegradation, the
experiment was stopped and the biomass used for cultivation of SMX biodegrading organisms on solid R2A-UV media (1.5% agar supply). SMX
removal was determined by UV-absorbance measurements (UV-AM) as fast
pre-screening method for biodegradation (see 2.4.1).

6.4.3.2 CULTIVATION AND ISOLATION OF PURE CULTURES

Pure cultures were successfully cultivated and isolated from SMX-acclimated biodegrading ASC. 200 µL AS was plated on solid R2A-UV media
containing 10 mg L^{-1} SMX to inhibit growth of non-resistant bacteria and
foster growth of potential SMX-resistant/biodegrading organisms. After
cultures were observed on solid media they were isolated and further purified by streaking on new plates resulting in 110 isolates. These were used
for inoculation of 100 mL setups with 20 mL MSM-CN media (KH$_2$PO$_4$ 80
mg L^{-1}, K$_2$HPO$_4$ 200 mg L^{-1}, Na$_2$HPO$_4$ 300 mg L^{-1}, MgSO$_4$*7 H$_2$O 20 mg

L^{-1}, $CaCl*2$ H_2O 40 mg L^{-1}, $FeCl_3*6$ H_2O 0.3 mg L^{-1}, sodium acetate 300 mg L^{-1} and NH_4NO_3 7.5 mg L^{-1}, DOC:N ratio 33:1, pH 7.4) spiked with 10 mg L^{-1} SMX. Setups were monitored with UV-AM (see 2.4.1) for possible biodegradation. Isolates showing biodegradation were further identified by 16S rRNA gene sequence analysis (see 2.5).

6.4.3.3 BIODEGRADATION SETUPS WITH PURE CULTURES

Batch experiments were performed to A) screen for biodegradation potential in the isolated cultures and B) determine differences in SMX biodegradation pattern and rate concerning the availability of nutrients. Three media, R2A-UV, MSM-CN and MSM (as MSM-CN but without sodium acetate and NH_4NO_3) were used and inoculated with pure cultures in 100 mL setups filled with 20 mL of media spiked with 10 mg L^{-1} SMX. Duplicate setups (n=2) including sterile, i.e. autoclaved biomass and abiotic, i.e. without biomass, controls for each medium were prepared. Aerobic conditions and photolysis prevention were ensured by shaking at 150 rpm on an orbital shaker in the dark.

The setups were sampled once a day for MSM-CN and MSM media and twice a day for R2A-UV, by taking 1 mL supernatant after half an hour of sedimentation that was sufficient to ensure not to withdraw much biomass. 200 μL was used for UV-AM and 800 μL for LC-UV measurements.

6.4.4 ANALYSES OF SULFAMETHOXAZOLE

6.4.4.1 UV-AM

200 μL were taken from the setups and directly used for UV-AM as described elsewhere (Herzog et al., submitted) with the following changes applied. Calibration was performed with 1.0, 5.0, 10.0 and 15.0 mg L^{-1} SMX in high-purity water and the used media to evaluate measurement reliability

and background absorbance. 96 well UV-star plates from Greiner Bio-One (Greiner Bio-One GmbH, Frickenhausen, Germany) filled with 200 µL were used for measurements and analyzed with an automated plate reader (EnSpire® Multimode Plate Reader, Perkin Elmer, Rodgau, Germany). Each measurement included an SMX blank (media with SMX but without organisms) was measured to detect changes over time as well as a blank (media without SMX) to detect background absorbance.

6.4.4.2 LC-UV ANALYSIS

800 µL samples obtained from the setups were centrifuged (10 min, 8000 g, 20°C), filtrated through a 0.45 µm membrane filter to remove cellular debris and biomass and filled into sterile glass flasks. Flasks were stored at-20°C before analysis.

Analysis was performed with a Dionex 3000 series HPLC system (Dionex, Idstein, Germany), equipped with an auto sampler. A DAD scanning from 200 to 600 nm was applied to detect and quantify SMX. Chromatographic separation was achieved on a Nucleosil 120-3 C18 column (250 mm × 3.0 mm i.d., 3 µm particle size) from Macherey Nagel (Düren, Germany) at a column temperature of 25°C. The applied mobile phases were acetonitrile (AN) and water (pH 2.5 using phosphoric acid). The gradient used for the first 5 min was 7% AN followed by 7-30% AN from 5-18 min, 30% AN for minutes 18-30 and finally 7% AN for minutes 30-35. The solvent flow rate was 0.6 mL min^{-1}. The column was allowed to equilibrate for 5 min between injections. Limit of quantification and limit of detection were 0.1 mg L^{-1} and 0.03 mg L^{-1}, respectively.

6.4.5 TAXONOMIC AND PHYLOGENETIC IDENTIFICATION OF ISOLATED PURE CULTURES BY 16S RRNA GENE SEQUENCE ANALYSIS

DNA of SMX biodegrading organisms was extracted by a standard phenol/chloroform/CTAB extraction method. 16S rRNA gene was subsequently amplified via standard PCR using universal bacterial primers 27f (5-AGA

GTT TGA TCM TGG CTC AG-3) and 1492r (5-TAC GGY TAC CTT GTT ACG ACT T-3) [49]. All cultures were sent to MWG Operon (Ebersberg, Germany) for sequencing using again primers 27f and 1492r and resulting in nearly full length 16S rRNA gene sequences. Sequences were analyzed with and submitted to European Nucleotide Archive (http://www.ebi.ac.uk/ena/) to receive accession numbers (Table 2).

Subsequent phylogenetic analysis was accomplished with the sequences using the alignment and tree calculation methods of the ARB software package [50]. The nearly complete 16S rRNA gene sequences of the species isolated in this study and their corresponding published closest relatives (http://blast.ncbi.nlm.nih.gov/Blast.cgi) were added to an existing ARB-alignment for the 16S rRNA gene sequence. Alignment was performed with the CLUSTAL W implemented in ARB. Phylogenetic trees of the 16S rRNA gene sequences were calculated based on maximum likelihood.

REFERENCES

1. Kümmerer K: Pharmaceuticals in the environment: sources, fate, effects, and risks. 2nd edition. Berlin, Heidelberg, Germany: Springer; 2004.
2. Kümmerer K: Pharmaceuticals in the environment. 3rd, Revised and enlarged Edition edn. Berlin, Heidelberg, Germany: Springer; 2008.
3. Baran W, Sochacka J, Wardas W: Toxicity and biodegradability of sulfonamides and products of their photocatalytic degradation in aqueous solutions. Chemosphere 2006, 65:1295–1299.
4. Xu B, Mao D, Luo Y, Xu L: Sulfamethoxazole biodegradation and biotransformation in the water-sediment system of a natural river. Bioresour Technol 2011, 102:7069–7076.
5. Heberer T: Occurrence, fate, and removal of pharmaceutical residues in the aquatic environment: a review of recent research data. Toxicol Lett 2002, 131:5–17.
6. Ternes T, Joss A: Human pharmaceuticals, hormones and fragrances the challenge of micropollutants in urban water management. ; 2007.
7. Kümmerer K: Antibiotics in the aquatic environment-a review-part I. Chemosphere 2009, 75:417–434.
8. Kümmerer K: Antibiotics in the aquatic environment a review-part II. Chemosphere 2009, 75:435–441.
9. Pérez S, Eichhorn P, Aga DS: Evaluating the biodegradability of sulfamethazine, sulfamethoxazole, sulfathiazole, and trimethoprim at different stages of sewage treatment. Environ Toxicol Chem 2005, 24:1361–1367.

10. Hoa PTP, Managaki S, Nakada N, Takada H, Shimizu A, Anh DH, Viet PH, Suzuki S: Antibiotic contamination and occurrence of antibiotic-resistant bacteria in aquatic environments of northern Vietnam. Sci Total Environ 2011, 409:2894–2901.

11. Agerso Y, Petersen A: The tetracycline resistance determinant Tet 39 and the sulphon-amide resistance gene sulII are common among resistant *Acinetobacter* spp. isolated from integrated fish farms in Thailand. J Antimicrob Chemother 2007, 59:23–27.

12. Szczepanowski R, Linke B, Krahn I, Gartemann K-H, Gützkow T, Eichler W, Pühler A, Schlüter A: Detection of 140 clinically relevant antibiotic-resistance genes in the plasmid metagenome of wastewater treatment plant bacteria showing reduced sus-ceptibility to selected antibiotics. Microbiology 2009, 155:2306–2319.

13. Cavallucci S: Top 200: What's topping the charts in prescription drugs this year, Pharmacy practice, Canadian Healthcare Network. 2007.

14. Benotti MJ, Trenholm RA, Vanderford BJ, Holady JC, Stanford BD, Snyder SA: Pharmaceuticals and endocrine disrupting compounds in US drinking water. Envi-ron Sci Technol 2008, 43:597–603.

15. Miège C, Choubert J, Ribeiro L, Eusèbe M, Coquery M: Fate of pharmaceuticals and personal care products in wastewater treatment plants-Conception of a database and first results. Environ Pollut 2009, 157:1721–1726.

16. Sacher F, Lange FT, Brauch HJ, Blankenhorn I: Pharmaceuticals in groundwaters: analytical methods and results of a monitoring program in Baden-Wurttemberg, Germany. J Chromatogr 2001, 938:199–210.

17. Onesios K, Yu J, Bouwer E: Biodegradation and removal of pharmaceuticals and personal care products in treatment systems: a review. Biodegradation 2009, 20:441–466.

18. Huang T-S, Kunin CM, Yan B-S, Chen Y-S, Lee SS-J, Syu W: Susceptibility of My-cobacterium tuberculosis to sulfamethoxazole, trimethoprim and their combination over a 12 year period in Taiwan. J Antimicrob Chemother 2012, 67:633–637.

19. Fajardo A, Martínez JL: Antibiotics as signals that trigger specific bacterial respons-es. Curr Opin Microbiol 2008, 11:161–167.

20. Jiang X, Shi L: Distribution of tetracycline and trimethoprim/sulfamethoxazole re-sistance genes in aerobic bacteria isolated from cooked meat products in Guang-zhou, China. Food Control 2013, 30:30–34.

21. Liu F, Wu J, Ying G-G, Luo Z, Feng H: Changes in functional diversity of soil microbial community with addition of antibiotics sulfamethoxazole and chlortetra-cycline. Appl Microbiol Biotechnol 2012, 95:1615–1623.

22. Gutiérrez I, Watanabe N, Harter T, Glaser B, Radke M: Effect of sulfonamide antibi-otics on microbial diversity and activity in a Californian Mollic Haploxeralf. J Soils Sed 2010, 10:537–544.

23. Collado N, Buttiglieri G, Marti E, Ferrando-Climent L, Rodriguez-Mozaz S, Bar-celó D, Comas J, Rodriguez-Roda I: Effects on activated sludge bacterial commu-nity exposed to sulfamethoxazole. Chemosphere 2013, 93:99–106.

24. Göbel A, McArdell CS, Joss A, Siegrist H, Giger W: Fate of sulfonamides, macro-lides, and trimethoprim in different wastewater treatment technologies. Sci Total Environ 2007, 372:361–371.

25. Niu J, Zhang L, Li Y, Zhao J, Lv S, Xiao K: Effects of environmental factors on sulfamethoxazole photodegradation under simulated sunlight irradiation: kinetics and mechanism. J Environ Sci 2013, 25:1098–1106.

26. Trovó AG, Nogueira RFP, Agüera A, Sirtori C, Fernández-Alba AR: Photodegradation of sulfamethoxazole in various aqueous media: persistence, toxicity and photoproducts assessment. Chemosphere 2009, 77:1292–1298.

27. Hyland KC, Dickenson ERV, Drewes JE, Higgins CP: Sorption of ionized and neutral emerging trace organic compounds onto activated sludge from different wastewater treatment configurations. Water Res 2012, 46:1958–1968.

28. Ludwig W, Klenk H-P: Overview: a phylogenetic backbone and taxonomic framework for procaryotic systematics. In Bergey's manual® of systematic bacteriology. Edited by Boone D, Castenholz R. New York: Springer; 2001:49–65.

29. Bouju H, Ricken B, Beffa T, Corvini PF, Kolvenbach BA: Isolation of bacterial strains capable of sulfamethoxazole mineralization from an acclimated membrane bioreactor. Appl Environ Microbiol 2012, 78:277–279.

30. Jiang Q, Fu B, Chen Y, Wang Y, Liu H: Quantification of viable but nonculturable bacterial pathogens in anaerobic digested sludge. Appl Microbiol Biotechnol 2013, 97:6043–6050.

31. Wagner M, Assmus B, Hartmann A, Hutzler P, Amann R: In situ analysis of microbial consortia in activated sludge using fluorescently labelled, rRNA-targeted oligonucleotide probes and confocal scanning laser microscopy. J Microsc 1994, 176:181–187.

32. Vartoukian SR, Palmer RM, Wade WG: Strategies for culture of 'unculturable' bacteria. FEMS Microbiol Lett 2010, 309:1–7.

33. Wagner M, Amann R, Lemmer H, Schleifer KH: Probing activated sludge with oligonucleotides specific for proteobacteria: inadequacy of culture-dependent methods for describing microbial community structure. Appl Environ Microbiol 1993, 59:1520–1525.

34. Snaidr J, Amann R, Huber I, Ludwig W, Schleifer KH: Phylogenetic analysis and in situ identification of bacteria in activated sludge. Appl Environ Microbiol 1997, 63:2884–2896.

35. Reasoner DJ, Geldreich EE: A new medium for the enumeration and subculture of bacteria from potable water. Appl Environ Microbiol 1985, 49:1–7.

36. Chiellini C, Munz G, Petroni G, Lubello C, Mori G, Verni F, Vannini C: Characterization and comparison of bacterial communities selected in conventional activated sludge and membrane bioreactor pilot plants: a focus on nitrospira and planctomycetes bacterial phyla. Curr Microbiol 2013, 67:77–90.

37. Wells GF, Park H-D, Eggleston B, Francis CA, Criddle CS: Fine-scale bacterial community dynamics and the taxa-time relationship within a full-scale activated sludge bioreactor. Water Res 2011, 45:5476–5488.

38. Larcher S, Yargeau V: Biodegradation of sulfamethoxazole by individual and mixed bacteria. Appl Microbiol Biotechnol 2011, 91:211–218.

39. De Gusseme B, Vanhaecke L, Verstraete W, Boon N: Degradation of acetaminophen by *delftia tsuruhatensis* and *pseudomonas aeruginosa* in a membrane bioreactor. Water Res 2011, 45:1829–1837.

40. Tezel U, Tandukar M, Martinez RJ, Sobecky PA, Pavlostathis SG: Aerobic biotrans-formation of n-tetradecylbenzyldimethylammonium chloride by an enriched *pseudomonas* spp. Community. Environ Sci Technol 2012, 46:8714–8722.

41. Shiomi N, Ako M: Biodegradation of melamine and cyanuric acid by a newly-isolated *microbacterium* strain. Adv Microbiol 2012, 2:303–309.

42. Chunming W, Chunlian LIDW: Biodegradation of naphthalene, phenanthrene, anthracene and pyrene by *microbacterium* sp. 3-28. Chin J Appl Environ Biol 2009, 3:017.

43. Satola B, Wübbeler J, Steinbüchel A: Metabolic characteristics of the species *variovorax paradoxus*. Appl Microbiol Biotechnol 2013, 97:541–560.

44. Islas-Espinoza M, Reid B, Wexler M, Bond P: Soil bacterial consortia and previous exposure enhance the biodegradation of sulfonamides from Pig manure. Microb Ecol 2012, 64:140–151.

45. Gauthier H, Yargeau V, Cooper DG: Biodegradation of pharmaceuticals by *rhodococcus rhodochrous* and *aspergillus niger* by co-metabolism. Sci Total Environ 2010, 408:1701–1706.

46. Cohen GN: Bacterial growth. In Microbial biochemistry. Dordrech, Netherlands: Springer; 2011:1–10.

47. Yang S-F, Lin C-F, Wu C-J, Ng K-K, Yu-Chen Lin A, Andy Hong P-K: Fate of sulfonamide antibiotics in contact with activated sludge-sorption and biodegradation. Water Res 2012, 46:1301–1308.

48. Müller E, Schüssler W, Horn H, Lemmer H: Aerobic biodegradation of the sulfonamide antibiotic sulfamethoxazole by activated sludge applied as co-substrate and sole carbon and nitrogen source. Chemosphere 2013, 92:969–978.

49. Weisburg WG, Barns SM, Pelletier DA, Lane DJ: 16S ribosomal DNA amplification for phylogenetic study. J Bacteriol 1991, 173:697–703.

50. Ludwig W, Strunk O, Westram R, Richter L, Meier H, Buchner A, Lai T, Steppi S, Jobb G, Yadhukumar, et al: ARB: a software environment for sequence data. Nucleic Acids Res 2004, 32:1363–1371. doi:10.1186/1471-2180-13-276

CHAPTER 7

Assessing Bacterial Diversity in a Seawater-Processing Wastewater Treatment Plant by 454-Pyrosequencing of the 16S rRNA and *amoA* Genes

OLGA SÁNCHEZ, ISABEL FERRERA, JOSE M. GONZÁLEZ, AND JORDI MAS

7.1 INTRODUCTION

Activated sludge constitutes a crucial tool in the biodegradation of organic materials, transformation of toxic compounds into harmless products and nutrient removal in wastewater treatment plants (WWTPs). It contains a highly complex mixture of microbial populations whose composition has been intensively studied in the past decades. By applying culture-dependent methods many species have been isolated from activated sludge (Dias and Bhat, 1964; Prakasam and Dondero, 1967; Benedict and Carlson, 1971). However, a great majority cannot be obtained by conventional techniques (Wagner et al., 1993) and, consequently, current molecular

Assessing Bacterial Diversity in a Seawater-Processing Wastewater Treatment Plant by 454-Pyrosequencing of the 16S rRNA and amoA Genes. © *Sánchez O, Ferrera I, González J M, Mas J.* Microbial Biotechology 6,4 (2013), doi:10.1111/1751-7915.12052. *Used with the authors' permission.*

techniques such as sequence analysis of 16S rRNA gene clone libraries (Snaidr et al., 1997), fingerprinting methods such as denaturing gradient gel electrophoresis (DGGE; Boon et al., 2002), thermal gradient gel electrophoresis (TGGE; Eichner et al., 1999) and terminal restriction fragment length polymorphism (Saikaly et al., 2005) along with fluorescence in situ hybridization (FISH) have been employed in wastewater microbiology to analyse and compare the microbial structure of activated sludge. Recently, PCR-based 454 pyrosequencing has been applied to investigate the microbial populations of activated sludge in different WWTPs as well as in full-scale bioreactors (Sanapareddy et al., 2009; Kwon et al., 2010; Kim et al., 2011; Ye et al., 2011; Zhang et al., 2011a; b), greatly expanding our knowledge on activated sludge biodiversity.

An important process in WWTPs is nitrification, in which ammonium is removed by converting it first into nitrite and then to nitrate. Different bacterial species involved in this process have been characterized by means of clone library analysis in addition to FISH (Juretschko et al., 1998; Purkhold et al., 2000; Daims et al., 2001; Zhang etal., 2011b). Several ammonia-oxidizing and nitrite-oxidizing bacterial populations belonging to the phylum *Nitrospira* and to *Beta-* and *Gammaproteobacteria* have been identified as key members in this process, such as the genera *Nitrosomonas*, *Nitrobacter*, *Nitrospira* and *Nitrosococcus* (Wagner et al., 2002; Zhang et al., 2011b).

Nevertheless, most studies of microbial diversity in WWTPs refer to freshwater plants, either domestic or industrial, and yet very little is known about plants that utilize seawater for their operation, mainly because there are still very few of these running in the world. Their utilization responds to the deficiency in hydric resources prevailing in their locations and their use will probably increase in the near future due to water shortage associated to global warming as many areas are experiencing today (Barnett et al., 2005). As a consequence, knowledge of the microbial diversity becomes crucial to identify the key players in these systems.

In a recent survey (Sánchez et al., 2011), the prokaryotic diversity of a seawater-utilizing WWTP from a pharmaceutical industry located in the south of Spain was characterized using a polyphasic approach by means of three molecular tools that targeted the 16S rRNA gene, i.e. DGGE, clone libraries and FISH. The results showed that the composition of the bacterial community differed substantially from other WWTP previously reported, since *Betaproteobacteria* did not seem to be the predominant group; in contrast, other classes of *Proteobacteria*, such as *Alpha-* and *Gammaproteobacteria*, as well as members of *Bacteroidetes* and *Deinoccocus-Thermus* contributed in higher proportions. Besides, utilization of specific primers for amplification of the *amoA* (ammonia monooxygenase subunit A) gene confirmed the presence of nitrifiers corresponding to the *Beta-* subclass of *Proteobacteria*, although they were not identified in this study.

In the present article, we further investigated the diversity of this system by applying 454-pyrosequencing, a much more powerful molecular technique, which provides thousands of sequence reads. We analysed the bacterial assemblage by targeting the 16S rRNA gene and increased our knowledge on its diversity by one order of magnitude. Additionally, we characterized the nitrifying members of this sludge by pyrosequencing the *amoA* gene. As far was we know, this is the first study that analyses the *amoA* gene diversity in an activated sludge of a WWTPs with the particularity to utilize seawater.

7.2 RESULTS AND DISCUSSION

We investigated the bacterial community structure and identified the nitrifying members from the activated sludge of a seawater-utilizing WWTP located in Almeria (Southeast Spain). The plant treats wastewater from a pharmaceutical industry. The mean influent flow of the plant is 300 m^3 h^{-1} and has a treatment volume of 32 000 m^3. Nitrogen and chemical oxygen demand sludge loads were about 150–170 kg h^{-1} and 900–1000 kg h^{-1} respectively. DNA was extracted from samples of aerated mixed activated sludge collected in 2 consecutive years (2007 and 2008).

7.2.1 DIVERSITY OF BACTERIAL COMMUNITIES IN ACTIVATED SLUDGE

After a rigorous quality control (see Experimental procedures and Table S1) a total of 16 176 16S rRNA tag sequences of sufficient quality were analysed (8010 sequences corresponding to year 2007 and 8166 sequences to year 2008) and grouped into operational taxonomic units using uclust at 3% cut-off level. The clustering resulted in a total of 320 different OTUs from which 107 were shared between samples (33.3%) as shown in the Venn diagram (Fig. 1A). The number of OTUs in 2007 was 201 and in 2008 was 226. Although the proportion of shared OTUs is rather low, the unique diversity in each sample corresponded mainly to rare OTUs (relative abun- dance below 1%). In the case of year 2007, the unique clones to that sample represented a 19% of the total reads, from which only three OTUs were above 1%. For the 2008 sample, the unique clones represented a 9% of total sequences and only two OTUs presented an abundance above 1%. These results are in agreement with previous observations in which DGGE analysis from both samples showed virtually the same pattern for universal primers amplifying Bacteria, suggesting that activated sludge was at a steady state at least for the most abundant phylotypes (Sánchez et al., 2011).

Richness was computed by the Chao1 estimator and analysis by rarefaction showed that the diversity in the two samples was within the same range, although slightly higher in 2008. However, we found that this depth of sequencing was not sufficient to saturate the curve and therefore, the actual diversity is likely much higher (Fig. 2). Nevertheless, if compared with the rarefaction curve from a clone library performed from the 2007 activated sludge sample, we observe that, by applying pyrosequencing we increased our knowledge on the diversity present by one order of magnitude. Rank-abundance curves (Fig. S1A) show that there were only a few abundant phylotypes and a long tail of rare taxa, therefore, most of the unknown diversity probably corresponds to rare diversity (Pedrós-Alió, 2007).

RDP Classifier was used to assign the representative OTU sequences into different phylogenetic bacterial taxa. Figure S2 shows the relative abundances of the different groups at the phylum and class level for both years. *Deinococcus-Thermus*, *Proteobacteria*, *Chloroflexi* and *Bacteroidetes* were abundant in both samples. Comparison with a previous survey (Sánchez et al., 2011) indicates that most of these groups were also retrieved by different molecular methodologies. However, the contribution of each group varied depending on the technique used. The bacterial clone library over-represented the *Deinococcus-Thermus* group, while the

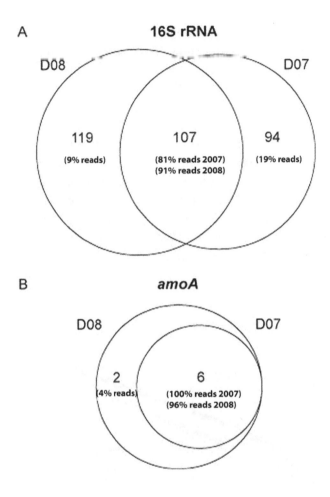

Figure 1: Venn diagrams of shared OTUs between the two samples (D07 and D08) for (A) 16S rRNA gene and (B) *amoA* gene. The number of reads that the OTUs represent is indicated in brackets.

rest of procedures showed similar results concerning this phylum. In contrast, the *Alphaproteobacteria* were over-represented by FISH (Fig. S3). (On the other hand, pyrosequencing allowed the detection of other groups that could not be recognized by other molecular techniques, such as the *Chloroflexi, Chlorobi, Deferribacteres, Verrumicrobia, Planctomycetes* and *Spirochaetes*, deepening our knowledge on the diversity of this activated sludge. Also, a certain percentage of sequences remained as unidentified bacteria (6.5% and 10.5% for years 2007 and 2008; Fig. S2). Except the *Chlorobi* and *Deferribacteres*, different pyrosequencing studies have reported the presence of these groups in conventional activated sludge samples (Sanapareddy et al., 2009; Kwon et al., 2010). However, it is remarkable that, in general, the proportions of the different groups in freshwater activated sludge were different from saline samples, and when going deeper into genus composition, the assemblage of our samples differs strongly from that previously reported. In general, prior pyrosequencing studies with different samples of activated sludge are in agreement with the predominance of the classes *Beta-* and *Gammaproteobacteria* and the phylum *Bacteroidetes* (Sanapareddy et al., 2009; Kwon et al., 2010), while

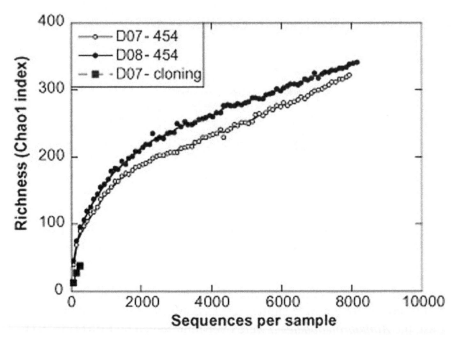

Figure 2: Rarefaction curves of 16S rRNA OTUs defined by 3% sequence variations in the activated sludge. Curves refer to the pyrosequence analyses of the 2 consecutive years (D07 and D08- 454) compared with the clone library analysis of the 2007 sludge (D07 – cloning).

in our saline activated sludge the groups that predominate are, within the phylum *Proteobacteria*, the *Alpha-* (8.0% and 7.3% in samples 2007 and 2008 respectively), *Gamma-* (19.0% and 21.4%) and *Deltaproteobacteria* (15.2% and 7.2%), as well as the *Deinococcus-Thermus* group (21.8% and 10.9%) and members of the phyla *Chloroflexi* (9.5% and 35.1%) and *Bacteroidetes* (18.3% and 2.8%). In contrast, Ye and colleagues (2011), who analysed by pyrosequencing the bacterial composition of a slightly saline activated sludge from a laboratory-scale nitrification reactor and a WWTP from Hong Kong, found that, in addition to *Proteobacteria* and *Bacteroidetes*, the phylum *Firmicutes* was also abundant in their samples; they also obtained similar groups as in the present study, such as the *Actinobacteria*, *Planctomycetes*, *Verrumicrobia*, *Deinococcus-Thermus*, *Chloroflexi* and *Spirochaetes*, although at different relative ratios, as well as different phyla not retrieved in the present work, for example the *Nitrospira*, *Chlamydiae* and TM7. Probably, differences are due to the feeding wastewater, since in our case the main influent corresponds to intermediate products of amoxicillin synthesis whereas in the other study the WWTP treated a slightly saline urban sewage from Hong Kong.

As in previous pyrosequencing studies (Keijser et al., 2008; Liu et al., 2008; Claesson et al., 2009), a part of sequences could only be assigned to the phylum/class level and the majority of taxa could not be classified at the genus level (74% for 2007 and 83.5% for 2008), demonstrating the extraordinary microbial diversity of activated sludge that cannot be classified using public 16S rRNA databases. Table S2 shows the taxa found in each sample in this study at the genus level, which are different from other reported genus of freshwater or either slightly saline activated sludge studies (Sanapareddy et al., 2009; Kwon et al., 2010; Ye et al., 2011). One of the most abundant genus in both samples is *Truepera*, a member of the phylum *Deinococcus-Thermus*, which includes radiation-resistant and thermophilic species. Although this phylum was also detected by pyrosequencing in a recent study with a slightly saline activated sludge (Ye et al., 2011), it accounted for no more than 0.6% of total community, while we found a significant percentage of sequences from this genus (21.8% and 10.9% for years 2007 and 2008 respectively).

7.2.2 DIVERSITY OF NITRIFYING COMMUNITY IN ACTIVATED SLUDGE

A total of 43 297 *amoA* gene sequences of good quality (11 236 reads for year 2007 and 32 061 reads for year 2008) were grouped into operational taxonomic units using uclust at 6% cut-off level. We selected a 6% cut-off to group closely related phylotypes of the *amoA* gene without losing potentially valuable information by the inclusion of phylogenetically distinct sequences. Interestingly, the diversity of the nitrifying bacterial community revealed by pyrosequencing of the *amoA* gene was very low and rarefaction analyses showed the depth of sequencing was sufficient to saturate the curve and recover the great majority. The clustering of 43 297 reads resulted in a total of only eight OTUs from which six were shared between samples as shown in the Venn diagram (Fig. 1B). The shared OTUs corresponded to 97% of total reads, which indicates that the nitrifying community was very similar both years.

All *amoA* sequences were highly related to previously described sequences in the GenBank database, both environmental and from isolates (Fig. 3). Phylogenetic analysis revealed that eight phylotypes formed two separate clusters. The first cluster, which contains three OTUs, was mostly retrieved in the 2008 library and represented 45.4% of sequences of that sample. The closest relatives in GenBank database (99% similarity) included sequences from organisms that have not been obtained from a WWTP, and *Nitrosomonas* sp. LT-2 and LT-5, isolated from a CANON reactor (98% identity). The second cluster, which contains five of the OTUs and represented the most abundant phylotypes in both samples, was most closely related (94% identity) to cultured representatives of strains of *Nitrosomonas marina* isolated from a biofilter of a recirculating shrimp aquaculture system (GenBank Accession No. HM345621, HM345612 and HM345618) and *Nitrosomonas* sp. NS20 isolated from coastal marine sediments. This cluster virtually represented all sequences (99.99%) in the sample of 2007 whereas in 2008 it comprised 54.6%.

These data are consistent with previous results found by Ye and colleagues (2011) in slightly saline activated sludge, which showed that *Nitrosomonas*, together with *Nitrospira*, was the dominant nitrifying genera, and also with the study by Park and colleagues (2009), who identified that a specific ammonia-oxidizing bacteria belonging to the *Nitrosomonas europaea* lineage was dominant in a full-scale bioreactor treating saline wastewater due to its adaptation to high-salt conditions. In general, nitrosomonads are also responsible for ammonia oxidation in conventional WWTPs (Purkhold et al., 2000; Zhang et al., 2011b). However, we did not retrieve *Nitrosomonas* in the pyrosequencing 16S rRNA libraries or previously in DGGE gels, clone libraries and FISH (Sánchez et al., 2011) probably due to their low abundance. Thus, pyrosequencing of functional genes such as *amoA* revealed the presence of particular groups which could not be retrieved when analysing the 16S rRNA, demonstrating its value to deepen into the functionality of microbial populations when targeting specific genes. The only nitrifier that could be retrieved in our samples by 16S rRNA pyrosequencing was *Nitrosococcus*, a *Gammaproteobacteria* which just represented 0.3% and 0.5% of the total reads for years 2007 and 2008 respectively, and has also been reported to be an important nitrifier in some activated sludges (Juretschko et al., 1998; Raszka et al., 2011). Thus, since it was actually detectable in the general bacterial 16S rRNA gene population, it could also participate in ammonia oxidization together with the *Betaproteobacteria*, despite previous efforts for amplifying the gammaproteobacterial *amoA* gene yielded negative results (Sánchez et al., 2011).

On the other hand, we know that nitrification and denitrification are central processes in our system, since a nitrification fraction of 98% and a total nitrogen removal over 80% have been reported (M.I. Maldonado, pers. comm.). In fact, Yu and Zhang (2012), when applying both metagenomic and metatranscriptomic approaches to characterize microbial structure and gene expression of an activated sludge community from a municipal WWTP in Hong Kong found that nitrifiers such as *Nitrosomonas* and *Nitrospira* had a high transcription activity despite presenting a very low abundance (they accounted only 0.11% and 0.02% respectively in their DNA data set), and the results from Zhang and colleagues (2011b) indicated that the abundance of ammonia-oxidizing bacteria in the activated sludge from different WWTPs was very low. Similarly, our results suggest that *Nitrosomonas* could be responsible of nitrification although showing a low abundance. Also, it may be possible that other genera different from the well-known *Betaproteobacteria* could contribute to nitrification activity.

In fact, different studies have reported the autotrophic oxidation of ammonia by members of the domain *Archaea*. The crenarchaeon *Nitrosopumilus maritimus* is able to oxidize ammonia to nitrite under mesophilic conditions (Könneke et al., 2005), while ammonia-oxidizing *Archaea* occurred in activated sludge bioreactors used to remove ammonia from wastewater (Park et al., 2006). However, amplification of the *amoA* gene to detect the presence of archaeal nitrifiers yielded negative results in our samples. In fact, *Archaea* accounted only for 6% of DAPI counts, and all sequences retrieved previously in an archaeal clone library were related to methanogenic archaea (Sánchez et al., 2011). Other studies have also demonstrated the presence of methanogens in aerated activated sludge but, although active, they played a minor role in carbon and nitrogen turnover (Gray et al., 2002; Fredriksson et al., 2012).

Interestingly, different heterotrophic bacteria, such as *Bacillus* sp. (Kim et al., 2005), *Alcaligenes faecalis* (Liu etal., 2012), *Marinobacter* sp. (Hai-Yan etal., 2012), *Achromobacter xylosoxidans* (Kundu et al., 2012) and *Pseudomonas* sp. (Su et al., 2006) have been described as potential nitrifiers, and remarkably, some of these genera have been isolated from our activated sludge by culture-dependent techniques (data not shown); for instance, some strains were identified as *Bacillus* sp., *Alcaligenes* sp., *Marinobacter hydrocarbonoclasticus* and *Pseudomonas* sp.

In contrast, sequences of *Nitratireductor* sp., a denitrifying microorganism, have been retrieved with different molecular methods (pyrosequencing in this study and DGGE and clone library in Sánchez et al., 2011), while other sequences from potential denitrifiers have been recovered only by 454-pyrosequencing, such as *Leucobacter* sp., *Caldithrix* sp., *Castellaniella* sp. and *Halomonas* sp. Besides, other candidates for denitrifying bacteria have been isolated by culture-dependent techniques, such as *Alcaligenes* sp., *Bacillus* sp., *Paracoccus* sp., *Pseudomonas* sp. and *Marinobacter* sp. (data not shown). Other genera retrieved by pyrosequencing were related to the nitrogen fixation process, that is, *Microbacterium* sp., *Aminobacter* sp. and *Spirochaeta* sp., while *Sphingomonas* was detected by clone library and culture-dependent methods.

Summarizing, we can conclude that the bacterial diversity in the activated sludge of the seawater-processing plant was high as previously observed in conventional WWTPs. However, the composition of the bacterial community differed strongly from other plants, and was dominated by *Deinococcus-Thermus*, *Proteobacteria*, *Chloroflexi* and *Bacteroidetes*. Previous analyses by clone library, DGGE and FISH were not enough to reflect the profile of the bacterial community in wastewater sludge and although pyrosequencing was a powerful tool to define the microbial composition deeper sequencing is required. Despite nitrification rates were high in the system, known ammonia-oxidizing bacteria were not identified by means of 16S rRNA studies and analysis of the specific functional gene *amoA* was required to reveal the presence and identity of the bacteria responsible for this process. These results suggest that only a few populations of low abundant but specialized bacteria likely with high transcription activity are responsible for removal of ammonia in these systems. However, further studies to isolate the key microorganisms involved in ammonia- oxidation will be essential in order to understand this process in saline WWTPs.

REFERENCES

1. Barnett, T.P., Adam, J.C., and Lettenmaier, D.P. (2005) Potential impacts of a warming climate on water availability in snow-dominated regions. Nature 438: 303–309.
2. Benedict, R.G., and Carlson, D.A. (1971) Aerobic heterotrophic bacteria in activated sludge. Water Res 5: 1023–1030.
3. Boon, N., De Windt, W., Verstraete, W., and Top, E.M. (2002) Evaluation of nested PCR-DGGE (denaturing gradient gel electrophoresis) with group-specific 16S rRNA primers for the analysis of bacterial communities from different wastewater treatment plants. FEMS Microbiol Ecol 39: 101–112.
4. Claesson, M.J., O'Sullivan, O., Wang, Q., Nikkilä, J., Marchesi, J.R., Smidt, H., et al. (2009) Comparative analysis of pyrosequencing and a phylogenetic microarray for exploring microbial community structures in the human distal intestine. PLoS ONE 4: e6669.
5. Daims, H., Nielsen, J.L., Nielsen, P.H., Schleifer, K.H., and Wagner, M. (2001) In situ characterization of Nitrospira-like nitrite-oxidation bacteria active in wastewater treatment plants. Appl Environ Microbiol 67: 5273–5284.

6. Dias, F.G., and Bhat, J.V. (1964) Microbial ecology of activated sludge. Appl Environ Microbiol 12: 412–417.
7. Eichner, C.A., Erb, R.W., Timmis, K.N., and Wagner-Döbler, I. (1999) Thermal gradient gel electrophoresis analysis of bioprotection from pollutant shocks in the activated sludge microbial community. Appl Environ Microbiol 65: 102–109.
8. Fredriksson, N.J., Hermansson, M., and Wilén, B.-M. (2012) Diversity and dynamics of *Archaea* in an activated sludge wastewater treatment plant. BMC Microbiol 12: 140.
9. Gray, N.D., Miskin, E.P., Kornilova, O., Curtis, T.P., and Head, I.M. (2002) Occurrence and activity of *Archaea* in aerated activated sludge wastewater treatment plants. Environ Microbiol 4: 158–168.
10. Hai-Yan, Z., Ying, L., Xi-Yan, G., Guo-Min, A., Li-Li, M., and Zhi-Pei, L. (2012) Characterization of a marine origin aerobic nitrifying–denitrifying bacterium. J Biosci Bioeng 114: 33–37.
11. Juretschko, S., Timmerman, G., Schmid, M., Schleifer, K.-H., Pommerening-Roser, A., Koops, H., and P., and Wagner, M. (1998) Combined molecular and conventional analyses of nitrifying bacterium diversity in activated sludge: *Nitrosococcus mobilis* and *Nitrospira*-like bacteria as dominant populations. Appl Environ Microbiol 64: 3042–3051.
12. Keijser, B.J., Zaura, E., Huse, S.M., van der Vossen, J.M., Schuren, F.H., Montijn, R.C., et al. (2008) Pyrosequencing analysis of the oral microflora of healthy adults. J Dent Res 87: 1016–1020.
13. Kim, J.K., Park, K.P., Cho, K.S., Nam, S.W., Park, T.J., and Bajpai, R. (2005) Aerobic nitrification-denitrification by heterotrophic *Bacillus* strains. Bioresour Technol 96: 1897–1906.
14. Kim, T.-S., Kim, H.-S., Kwon, S., and Park, H.-D. (2011) Nitrifying bacterial community structure of a full-scale integrated fixed-film activated sludge process as investigated by pyrosequencing. J Microbiol Biotechnol 21: 293–298.
15. Könneke, M., Bernhard, A.E., de la Torre, J.R., Walker, C.B., Waterbury, J.B., and Stahl, D.A. (2005) Isolation of an autotrophic ammonia-oxidizing marine archaeon. Nature 437: 543–546.
16. Kundu, P., Pramanik, A., Mitra, S., Choudhury, J.D., Mukher- jee, J., and Mukherjee, S. (2012) Heterotrophic nitrification by *Achromobacter xylosoxidans* S18 isolated from a small-scale slaughterhouse wastewater. Bioprocess Biosyst Eng 35: 721–728.
17. Kwon, S., Kim, T.-S., Yu, G.H., Jung, J.-H., and Park, H.-D. (2010) Bacterial community composition and diversity of a full-scale integrated fixed-film activated sludge system as investigated by pyrosequencing. J Microbiol Biotechnol 20: 1717–1723.
18. Liu, Y., Li, Y., and Lv, Y. (2012) Isolation and characterization of a heterotrophic nitrifier from coke plant wastewater. Water Sci Technol 65: 2084–2090.
19. Liu, Z., DeSantis, T.Z., Andersen, G.L., and Knight, R. (2008) Accurate taxonomy assignments from 16S rRNA sequences produced by highly parallel pyrosequencers. Nucleic Acids Res 36: e120.
20. Park, H.-D., Wells, G.F., Bae, H., Criddle, C.S., and Francis, C.A. (2006) Occurrence of ammonia-oxidizing archaea in wastewater treatment plant bioreactors. Appl Environ Microbiol 72: 5643–5647.

21. Park, H.-D., Lee, S.-Y., and Hwang, S. (2009) Redundancy analysis demonstration of the relevance of temperature to ammonia-oxidizing bacterial community compositions in a full-scale nitrifying bioreactor treating saline wastewater. J Microbiol Biotechnol 19: 346–350.

22. Pedrós-Alió, C. (2007) Ecology. Dipping into the rare biosphere. Science 315: 192–193.

23. Prakasam, T.B.S., and Dondero, N.C. (1967) Aerobic heterotrophic populations of sewage and activated sludge. I. Enumeration. Appl Environ Microbiol 15: 461–467.

24. Purkhold, U., Pommerening-Röser, A., Juretschko, S., Schmid, M.C., Koops, H.P., and Wagner, M. (2000) Phylogeny of all recognized species of ammonia oxidizers based on comparative 16S rRNA and amoA sequence analysis: implications for molecular diversity surveys. Appl Environ Microbiol 66: 5368–5382.

25. Raszka, A., Surmacz-Górska, J., Zabczynski, S., and Miksch, K. (2011) The population dynamics of nitrifiers in ammonium-rich systems. Water Environ Res 83: 2159–2169.

26. Saikaly, P.E., Stroot, P.G., and Oerther, D.B. (2005) Use of 16S rRNA gene terminal restriction fragment analysis to assess the impact of solids retention time on the bacterial diversity of activated sludge. Appl Environ Microbiol 71: 5814–5822.

27. Sanapareddy, N., Hamp, T.J., González, L.C., Hilger, H.A., Fodor, A.A., and Clinton, S.M. (2009) Molecular diversity of a North Carolina wastewater treatment plant as revealed by pyrosequencing. Appl Environ Microbiol 75: 1688–1696.

28. Sánchez, O., Garrido, L., Forn, I., Massana, R., Maldonado, M.I., and Mas, J. (2011) Molecular characterization of activated sludge from a seawater-processing wastewater treatment plant. Microb Biotechnol 4: 628–642.

29. Snaidr, J., Amann, R., Huber, I., Ludwig, W., and Schleifer, K.-H. (1997) Phylogenetic analysis and in situ identification of bacteria in activated sludge. Appl Environ Microbiol 63: 2884–2896.

30. Su, J.J., Yeh, K.S., and Tseng, P.W. (2006) A strain of Pseudomonas sp. Isolated from piggery wastewater treatment systems with heterotrophic nitrification capability in Taiwan. Curr Microbiol 53: 77–81.

31. Wagner, M., Amman, R., Lemmer, H., and Schleifer, K.-H. (1993) Probing activated sludge with oligonucleotides specific for proteobacteria: inadequacy of culture-dependent methods for describing microbial community structure. Appl Environ Microbiol 59: 1520–1525.

32. Wagner, M., Loy, A., Nogueira, R., Purkhold, U., Lee, N., and Daims, H. (2002) Microbial community composition and function in wastewater treatment plants. Antonie Van Leeuwenhoek 81: 665–680.

33. Ye, L., Shao, M.-S., Zhang, T., Tong, A.H.Y., and Lok, S. (2011) Analysis of the bacterial community in a laboratory-scale nitrification reactor and a wastewater treatment plant by 454-pyrosequencing. Water Res 45: 4390–4398.

34. Yu, K., and Zhang, T. (2012) Metagenomic and metatranscriptomic analysis of microbial community structure and gene expression of activated sludge. PLoS ONE 7: e38183. doi:10.1371/journal.pone.0038183.

35. Zhang, T., Shao, M.F., and Ye, L. (2011a) 454 pyrosequencing reveals bacterial diversity of activated sludge from 14 sewage treatment plants. ISME J 6: 1137–1147.

36. Zhang, T., Ye, L., Tong, A.H.Y., Shao, M.-F., and Lok, S. (2011b) Ammonia-oxidizing *archaea* and ammoniaoxidizing bacteria in six full-scale wastewater treatment bioreactors. Appl Microbiol Biotechnol 91: 1215–1225.

There are several supplemental files that are not available in this version of the article. To view this additional information, please use the citation on the first page of this chapter.

PART III

NITROGEN AND PHOSPHORUS REMOVAL

CHAPTER 8

An Efficient Process for Wastewater Treatment to Mitigate Free Nitrous Acid Generation and its Inhibition on Biological Phosphorus Removal

JIANWEI ZHAO, DONGBO WANG, XIAOMING LI, QI YANG, HONGBO CHEN, YU ZHONG, HONGXUE AN, AND GUANGMING ZENG

8.1 INTRODUCTION

Enhanced biological phosphorus removal is one important strategy to protect natural waters from eutrophication. It is usually achieved through culturing an activated sludge with alternating anaerobic and oxic conditions, by which polyphosphate accumulating organisms (PAOs), the microorganisms responsible for phosphorus removal in wastewater treatment plants (WWTPs), can be largely enriched. To gain deep understandings regarding this biological phosphorus removal regime, numerous studies

An Efficient Process for Wastewater Treatment to Mitigate Free Nitrous Acid Generation and its Inhibition on Biological Phosphorus Removal. © *Zhao J, Wang D, Li X, Yang Q, Chen H, Zhong Y, An H, and Zeng G.* Scientific Reports 5, 8602 (2015), doi:10.1038/srep08602 *Licensed under Creative Commons Attribution 4.0 International License, http://creativecommons.org/licenses/by/4.0/.*

have been made in the past two decades [1,2]. It is widely accepted that PAOs take up available carbon sources anaerobically and store them as poly-b-hydroxyalkanoates (PHAs), with the energy and reducing power mainly gained through polyphosphate cleavage and glycogen degradation, respectively. In the subsequent oxic phase, the stored PHAs are utilized for cell growth, glycogen replenishment, phosphorus uptake, and poly-phosphate accumulation [2–4]. This metabolism behavior is considered to provide a selective advantage to PAOs over other populations.

In general, biological phosphorus removal can be excellently achieved in well-defined laboratory experiments, with the high abundance of PAOs above 90% [5]. In real WWTPs, however, unpredictable failures due to lost or reduced activity of PAOs are often observed [2]. This is primarily because biological phosphorus removal in full-scale WWTPs usually occurs along with biological nitrogen removal, by which denitrifiers will compete with PAOs for the limited carbon sources available in wastewaters, the recycled mixtures will disturb the anaerobic circumstance, and some intermediates of nitrogen removal such as nitrite and free nitrous acid (FNA) will inhibit the metabolisms of PAOs. Among them, the effect of FNA on the metabolisms of PAOs has been drawn much attention recently owing to its strong inhibition on the activities of PAOs [5–8].

Nitrite is inevitably produced in substantial amounts during biological nitrogen removal. It was reported that nitrite concentration could accumulate up to 12.3–22.6 mg/L in domestic wastewater treatment processes [9]. Especially in some WWTPs that achieve nitrogen removal via the nitrite pathway, the accumulated concentration could reach up to 40 mg/L [10]. Previous researchers considered that the intermediate of nitrification and denitrification (i.e., nitrite) caused seriously inhibition on the metabolisms of PAOs, but recently there have been increasing evidences showing that FNA, the protonated form of nitrite, rather than nitrite is the actual inhibitor [5,11,12]. For example, it was reported that FNA could inhibit aerobic phosphorus uptake seriously at a low level of 0.5×10^{-3} mg HNO_2-N/L [11], and more than 1.5×10^{-3} mg HNO_2-N/L could result in the complete loss of aerobic phosphorus uptake [13]. Pijuan et al. [5] showed that the aerobic phosphorus uptake was inhibited by 50% when FNA reached 0.52×10^{-3} mg HNO_2-N/L. Anoxic phosphorus uptake was also significantly affected by FNA presence [14], and 0.02 mg HNO_2-N/L would cause the complete loss of anoxic phosphorus uptake [11]. Furthermore, anaerobic metabolisms of PAOs were also severely affected by FNA, and Ye et al. [8] demonstrated that FNA had an adverse effect on carbon source uptake even at 1.0×10^{-3} mg HNO_2-N/L. Due to the severe inhibition on PAOs caused by FNA and the massive quantity of wastewaters treated daily, any improvement for reducing FNA generation or mitigating its inhibition on PAOs in current methods should have tangible economic and ecological consequences.

Several strategies, such as activated sludge adaption, pH adjustment, temperature control, and the feed flow and mode optimization, have been recommended to minimize the inhibitory effect of FNA on PAOs [6,15]. Though previous researches have proposed these meaningful methods, the strategy for mitigating the generation of FNA and its inhibition on PAOs from the aspect of modifying wastewater treatment operation regime has never been reported before. In addition, some previously proposed strategies such as pH adjustment and temperature control are rarely or not practically applied in full-scale WWTPs, probably due to the associated costs of adding pH controlling agents or increasing constructions. Thus, the method obtained in terms of wastewater treatment process modification may provide an alternatively practical option for engineers.

Besides the widely accepted anaerobic/oxic (A/O) phosphorus removal regime, PAOs are verified to be also enriched readily in the oxic/extended-idle (O/EI) wastewater treatment regime [16–19]. The O/EI regime enriches PAOs via some specific metabolic reactions (e.g., a significant idle release of phosphate and a low idle production of PHAs) occurred in the extended-idle phase, which shows a different inducing mechanism from the classical A/O regime. It is also reported that when receiving the same level of nitrate, the transformations of metabolic intermediates (especially the accumulation of nitrite) in the O/EI regime are much lower affected than those in the A/O regime [16]. Thus, one method that might be used to decrease FNA generation or mitigate its inhibition on PAOs from the viewpoint of wastewater treatment regime, we think, is to develop a suitable biological nutrient removal (BNR) process based on the O/EI regime. Although several studies have been performed in terms of the O/EI phosphorus removal regime, the questions as to whether (and how) this regime can achieve good performances of simultaneous nitrogen and phosphorus removal remain unknown. Additionally, it is also unclear whether PAOs cultured in this O/EI based on BNR process can tolerate higher level of disturbances caused by nitrogen removal (e.g., FNA inhibition) than those cultured in the conventional BNR process.

The purpose of this paper is to report this efficient method for significantly mitigating the generation of FNA and its inhibition on PAOs. Firstly, a new BNR process is designed based on the recently exploited O/EI regime, and its feasibility of BNR is evaluated. Since the O/EI regime is a phosphorus removal process with low nitrogen removal performances (around 60%), an anoxic phase is inserted into the oxic phase to enhance nitrogen removal. Therefore, the new BNR process developed here is performed as the oxic/anoxic/oxic/extended-idle (O/A/O/EI) regime. Then, the performances of BNR and the abundances of PAOs between the new process and conventional four-step (i.e., anaerobic/oxic/anoxic/oxic, defined as A/O/A/O) BNR process are compared. Finally, the reasons for the new process showing higher abundance of PAOs are explored via the www.nature.com/scientificreports analysis of cyclic pH variation, nitrite accumulation, and changes of metabolic intermediates in PAO metabolisms.

8.2 RESULTS

8.2.1 BNR PERFORMANCES IN THE O/A/O/EI REACTOR DURING THE LONG-TERM OPERATION

The data of the effluent NH_4^+-N, NO_2^--N, NO_3^--N, and soluble orthophosphate (SOP) concentrations in the O/A/O/EI reactor during the long-term operation are illustrated in Figure 1. It can be seen that the concentrations of these nutrients in effluent decreased along with the acclimated time. After domestication for about 40 d, the effluent nutrient concentrations became stable. The effluent NH_4^+-N, NO_2^--N, NO_3^--N, and SOP concentrations during stable operation were respectively maintained among 2.67–3.49, 0.15–0.31, 0.83–1.25 and 0.30–0.52 mg/L, which indicated that the efficiencies of nitrogen and phosphorus removal in the O/A/O/EI reactor were above 91% and 96%, respectively. The long-term experimental data showed that BNR could be successfully achieved in the new O/A/O/EI process.

8.2.2 COMPARISON OF BNR PERFORMANCES BETWEEN THE O/A/O/EI AND A/O/A/O REACTORS

The BNR performances between the O/A/O/EI and A/O/A/O reactors during a 21-day stable operation are summarized in Table 1. It was found that nitrogen removal was not obviously affected by the different operation processes. Although the effluent NH_4^+-N, NO_2^--N, NO_3^--N in the O/A/O/EI reactor were slightly lower than those in the A/O/A/O reactor, the efficiency of nitrogen removal between the two reactors was very close. However, the effluent SOP concentration in the O/A/O/EI reactor were much lower than that in the A/O/A/O reactor (0.41±0.11 mg/L versus 2.70±0.18 mg/L), which thereby caused a much higher phosphorus removal efficiency (97±0.73% versus 82±1.2%). FISH quantification further showed that the abundances of PAOs and glycogen accumulating organisms (GAOs) were respectively accounted for 41±7% and 11±3% in the O/A/O/EI reactor while the corresponding data in the A/O/A/O reactor were 30±5% and 24±4%, respectively (Figure 2), which were consistent with the phosphorus removal efficiency shown in Table 1. The above results clearly displayed that by modifying wastewater treatment operation regime the abundance of PAOs and the efficiency of phosphorus removal could be improved.

Phosphorus removal test via chemical precipitation at different pH. Batch tests in the absence of activated sludge microorganisms were performed to figure out the effect of pH on chemical phosphorus removal (Table S1, supporting information). As shown in Table S1, negligible SOP removal via chemical precipitation was observed. Only 2% of SOP was removed via chemical precipitation even the pH was 8.5, which implied that phosphorus removal in this study was dominated by biological effect.

8.2.3 COMPARISON OF THE EFFECT OF DIFFERENT FNA LEVELS ON PAO METABOLISMS BETWEEN THE TWO REACTORS

The two BNR reactors were developed from different inducing mechanisms of biological phosphorus removal. The different inducing mecha-

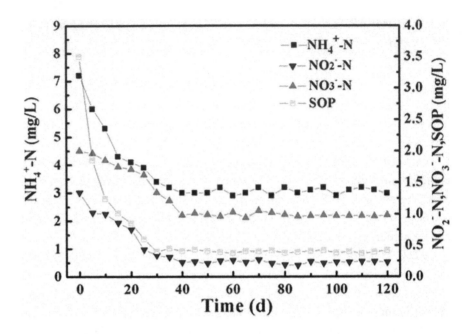

Figure 1: Variations of effluent NH_4^+-N, NO_2^--N, NO_3^--N, and SOP in O/A/O/EI reactor during the long-term operation.

nisms might give rise to different metabolic responses generated by PAOs even under the same level of FNA, thus we examined whether the same level of FNA would bring different effects on the metabolisms of PAOs between the two reactors. It can be seen from Table 2 that the metabolisms of PAOs in the conventional A/O/A/O reactor were severely inhibited by the FNA addition. When FNA concentration was 0, the effluent SOP in the A/O/A/O reactor was 0.75±0.05 mg/L. With the increased FNA concentration anaerobic SOP release, PHA-up/VFA ratio, Gly-de/VFA ratio, and Gly-syn were significantly decreased. As a result, effluent SOP concentration was increased largely. Especially when FNA was 0.51×10^{-3} mg/L, 7.05 ±0.09 mg/L of SOP was measured in the effluent, suggesting that only 53± 0.6% of influent SOP was removed. Also from Table 2, it can be found that the influence of FNA on the metabolisms of PAOs cultured in the O/A/O/ EI reactor was weaker than that in the conventional A/O/A/O reactor. Even

Table 1: Summary of reactor performances of the O/A/O/EI and A/O/A/O reactors during steady-state operation[a]

Item	O/A/O/EI reactor	A/O/A/O reactor
Effluent SOP (mg/L)	0.41±0.11	2.70±0.18
SOP removal efficiency (%)	97±0.73	82±1.2
Effluent NH_4^-N (mg/L)	2.68±0.41	3.75±0.15
Effluent NO_2^-N (mg/L)	0.23±0.08	0.31±0.03
Effluent NO_3^-N (mg/L)	0.85±0.21	1.21±0.09
Effluent pH	8.6±0.2	8.1±0.3
TN removal efficiency (%)	89±1.8	85±2.1

[a]Results are the average and standard deviation, and the data were obtained during the steady-state operation.

at FNA concentration of 0.51×10^{-3} mg/L, 5.10±0.08 mg/L was determined in the effluent, which indicated about 66±0.5% of influent SOP was removed. Further analysis revealed that compared with 0 mg/L of FNA, 0.51×10^{-3} mg/L of FNA caused by 33% decrease in phosphorus removal in the O/A/O/EI reactor, whereas the corresponding datum was 44.2% in the conventional A/O/A/O reactor. Similar observations were also observed in other FNA levels.

8.3 DISCUSSION

8.3.1 THE POSSIBLE MECHANISMS OF O/A/O/EI REGIME CULTURING HIGHER PAO ABUNDANCE

Several parameters can affect significantly the abundance of PAOs. By comparing the operational conditions between the two reactors, dissolved oxygen (DO) and pH might be the effect parameters since they are not constantly controlled during the whole process. Therefore, the cyclic variations of DO and pH between the two reactors during the steady-state operation were first compared, and the data are shown in Figure 3. Except for the anaerobic phase DO concentration in other phases of the A/O/A/O reactor showed very similar changes with that in the O/A/O/EI reactor. For example, in the experiment of day 80, DO in the O/A/O/EI reactor kept low levels during the initial period of first oxic phase and then gradually increased to a final concentration of 4.5 mg/L at the end of first oxic phase. During the subsequent anoxic phase, DO decreased rapidly to 0.7 mg/L and kept in the range of 0.4–0.7 mg/L in the remainder of anoxic phase. In the second oxic phase, DO increased gradually to 1.8 mg/L. After that, DO decreased gradually to 0.3 mg/L during the initial 60 min of idle phase and further decreased to 0.2 mg/L during the remainder of idle phase. Similar profiles were also made in other cycle studies. The results indicated that DO was not the main reason for the two reactors showing different PAO abundances.

The profile of pH change in the two reactors, however, exhibited obvious differences. In the O/A/O/EI reactor, pH gradually increased from 8.0 to 8.6 during the initial 60 min of first oxic phase and then decreased slightly during the remaining of this phase. In the following anoxic and oxic phases, a gradual increase to a final pH of 8.6 was observed. During the subsequent idle phase pH decreased gradually to the final pH of 8.2. In the A/O/A/O reactor, pH decreased from 8.0 to 7.4 in the anaerobic phase, and then a gradual increase followed by a slight decrease of pH was measured in the first oxic phase. In the subsequent anoxic phase, pH showed a gradual increase tendency then pH decreased slightly in the following oxic and idle phases. It can be clearly seen that cyclic variation of pH value in the O/A/O/EI reactor (8.0–8.6) was higher than that in the A/O/A/O reactor (7.2–8.2). The initial higher pH value achieved in the O/A/O/EI reactor might be mainly due to CO_2 expelled from the reactor by air-stripping, while the following pH decline was probably ascribed to nitrification. It was reported that denitrification and phosphorus uptake were the primarily reasons for pH increase in the anoxic phase and oxic phase [20]. During the extended-idle phase, a slight pH decline might be owing to idle SOP release (Figure 4). There are three forms of phosphorus existed in the activated sludge: metal phosphorus via physical chemistry processes,

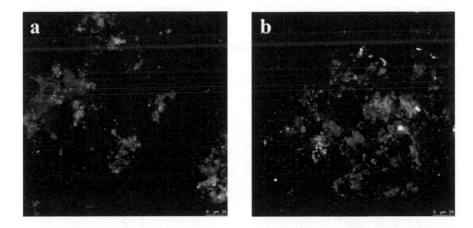

Figure 2: FISH micrographs of microbial communities from O/A/O/EI reactor (a) and A/O/A/O reactor (b) hybridizing with PAOmix (blue), GAOmix (red) and EUBmix(green) probes, respectively. Cells that were yellow had hybridized with both GAOmix and EUBmix probes. Samples were obtained after stable operation (on day 80).

intracellular polyphosphate inclusion, and bio-phosphorus for bacteria normal growth [21]. It is known that the amount of metal phosphorus in activated sludge is affected by pH, and higher pH value may cause higher chemical phosphorus removal [22]. However, batch tests showed that SOP removal via chemical precipitation was negligible (Table S1, supporting information), which implied SOP removal in this study was primarily due to biological effect. Previous publications showed that a high level of pH could provide a selective advantage to PAOs over other populations such as GAOs [19,23]. Accordingly, the higher level of cyclic pH was one reason for the O/A/O/EI reactor enriching more PAOs.

More importantly, the concentration of severe inhibitor to PAOs, i.e., FNA, is reported to closely relevant to pH value [24]. The different cyclic pH variations between the two reactors might cause different levels of FNA generation, thus the amount of FNA production between the two reactors was compared secondly. Besides pH, it is known that FNA concen-

Table 2: Effect of FNA on the phosphorus removal performance of O/A/O/EI and A/O/A/O reactors[a]

Influent FNA (x10⁻³ mg HNO_2-N/L)	PHA-up/VFA (C-mol/C-mol)	Gly-sin/VFA (mM-C/g VSS)	Effluent SOP (mg/L)	Idle SOP release (mg/L)	Anaerobic SOP release (mg/L)	Gly-de/VFA (C-mol/C-mol)	PHA-up/VFA (C-mol/C-mol)	Gly-syn (C-mol/g VSS)	Effluent SOP (mg/L)
0	0.52±0.03	0.51±0.04	0.22±0.03	6.89±0.04	62±0.7	0.62±0.04	1.19±0.02	1.87±0.09	0.75±0.05
0.05	0.48±0.05	0.49±0.05	0.39±0.05	6.22±0.06	58±1.2	0.61±0.03	1.06±0.05	1.72±0.05	1.04±0.07
0.15	042±0.02	0.47±0.08	1.15±0.07	5.79±0.03	53±0.9	0.59±0.05	1.01±0.04	1.59±0.07	3.85±0.09
0.26	0.38±0.05	0.41±0.04	1.55±0.11	5.42±0.06	49±0.8	0.55±0.02	0.92±0.06	1.36±0.04	4.05±0.11
0.38	0.35±0.06	0.40±0.02	2.36±0.05	5.18±0.02	43±1.4	0.54±0.03	0.79±0.07	1.24±0.06	5.12±0.12
0.51	0.31±0.03	0.38±0.03	5.10±0.08	4.87±0.05	35±0.5	0.53±0.04	0.71±0.05	1.12±0.07	7.05±0.09

[a]**Results are** the average and standard deviation, and the data were obtained during the steady-state operatio

tration is also relevant to temperature and nitrite concentration. Temperature between the two reactors was the same (20 ± 0.5 °C), and the change of nitrite as well as ammonia, nitrate, and SOP in the two reactors is shown in Figure 4. In the first oxic phase of O/A/O/EI reactor, SOP release was observed during the initial 30 min before SOP was swiftly taken up probably due to the low level of DO concentration (Fig. 3a). NH_4^+-N concentration was quickly decreased while NO_2^--N and NO_3^--N were substantially accumulated, with NO_2^--N and NO_3^--N accumulation up to 6.4 and 6.1 mg/L, respectively. In the subsequent anoxic phase, SOP and NH_4^+-N concentrations decreased slightly whereas NO_2^--N and NO_3^--N were largely decreased, suggesting that denitrification occurred in this period. Then, after 30 min of oxic phase (i.e., the second oxic phase), SOP, NH_4^+-N, NO_2^--N, and NO_3^--N concentrations in the effluent were 0.40, 3.0, 0.23, 0.97 mg/L, respectively. As comparison, it can be observed that a substantial amount of SOP was released in the anaerobic phase of conventional A/O/A/O reactor, then SOP uptake, NH_4^+-N oxidation, and NO_x^--N accumulation took place concurrently. During the subsequent anoxic phase, nitrate and nitrite reductions were clearly measured. After 30 min of oxic phase (i.e., the second oxic phase), SOP, NH_4^+-N, NO_2^--N, and NO_3^--N concentration in the effluent were 2.7, 3.1, 0.29, 1.12 mg/ L, respectively. Those behaviors were similar to the observations in the previous publications [25,26].

It should be highlighted that the maximal nitrite accumulation in the O/A/O/EI reactor was lower than that in the conventional A/O/A/O reactor (6.4 versus 7.5 mg/L), though the two reactors had approximately same effluent nitrite concentration. In addition, it can be found that pH value at the time for the O/A/O/EI reactor achieving its maximal nitrite accumulation was higher than that for the A/O/A/O reactor (8.4–8.5 versus 7.9). According to the formula proposed by Anthonisen et al.[24], the maximal FNA concentration generated in the O/A/O/EI reactor was about 0.52×10^{-4} mg HNO_2-N/L whereas the corresponding datum was 0.24×10^{-3} mg HNO_2-N/L in the A/O/A/O reactor. The maximum FNA concentration in the A/O/A/O reactor was approximately 4.6-time higher than that in the O/A/O/EI reactor. Similar observations were also observed in other cycles. It was reported that aerobic SOP uptake was severely affected when FNA was 0.26×10^{-3} mg HNO_2-N/L [7]. Although the time with maximal FNA concentration was low and cyclic FNA level in the two reactors changed

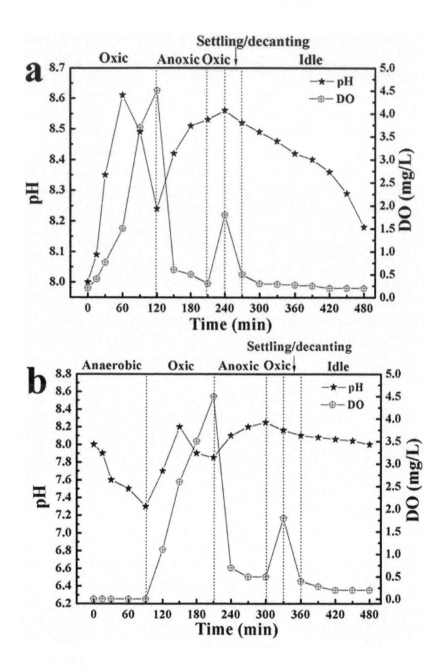

Figure 3: Variations of pH and DO in one typical cycle on Day 80 (a: O/ A/O/EI reactor; b: A/O/A/O reactor).

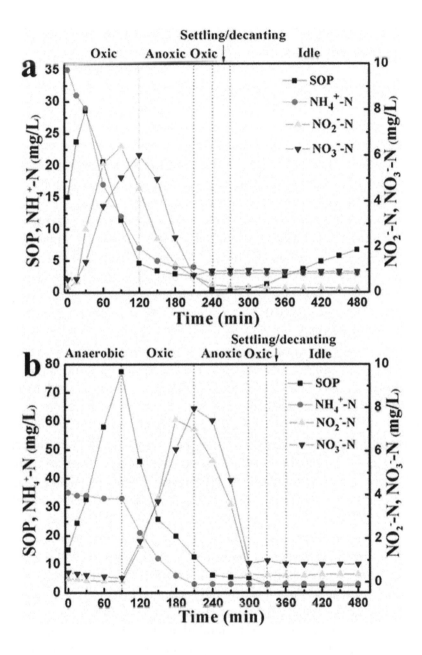

Figure 4: Changes of SOP, NH_4^+-N, NO_2^--N, and NO_3^--N in one typical cycle of O/A/O/EI (a) and A/O/A/O (b) reactors (on day 80).

with time, it could be found that the average FNA level in the O/A/O/EI reactor was lower than that in the A/O/A/O reactor. In addition, batch test showed that the O/A/O/EI reactor could alleviate the inhibition of FNA on the metabolisms of PAOs even under the same FNA level, as compared with the A/O/A/O reactor (Table 2). Therefore, it can be understood that the O/A/O/EI reactor enriched more PAOs than the conventional A/O/A/O reactor. Some scientists reported that PAOs could be acclimated high nitrite and FNA concentrations when using nitrite as sole electron acceptor [27]. However, O_2 prior to nitrite was the main electron acceptor in this study, which might be the reason for the inconsistent results.

FNA can inhibit or inactivate the activities of some key enzymes relevant to phosphorus removal. For instance, glyceraldehyde-3- phosphate dehydrogenase (GADP) and sulfhydryl (SH)-containing enzymes, which are respectively key enzymes involved in both glycolysis (gluconeogenesis) and the tricarboxylic acid (TCA) cycle, are reported to be heavily inhibited through reaction with FNA (Figure 5). The transformations of key metabolic intermediates such as glycogen and PHAs are closely related to glycolysis (gluconeogenesis) and the TCA cycle, thus the activity or abundance of PAOs will be reduced when FNA interferes with the pathways of glycolysis (gluconeogenesis) or the TCA cycle. From the "Methods" section, it can be found some differences between the two reactors. In the conventional A/O/A/O reactor, acetate is consumed in the anaerobic phase whereas it is taken up aerobically in the new O/A/O/EI reactor. This different metabolic behavior will cause certain metabolic differences, which might be one reason for the O/A/O/EI reactor enriching higher PAOs. Also, this different metabolic behavior might result in different metabolic responses of PAOs to FNA. For example, compared with the conventional A/O/A/O reactor where ATP and $NADH_2$ for PHAs formation were respectively provided via poly-P hydrolysis and glycogen degradation, the TCA cycle seems to supply both ATP and $NADH_2$ for PHAs synthesis in the O/A/O/EI reactor since it is generally accepted that the TCA cycle will dominate under aerobic conditions. However, it is still unclear why the O/A/O/EI reactor can alleviate the inhibition of FNA on the metabolisms of PAOs, as the TCA cycle plays an important role in PAO metabolisms of both regimes. Further efforts need to be carried out in future.

8.3.2 COMPARISON WITH OTHER STRATEGIES FOR MINIMIZING THE INHIBITORY EFFECT OF FNA ON PAOS

This paper presents an effective strategy for mitigating the generation of FNA and its inhibition on PAOs. That is, by modifying the wastewater treatment operation regime as the O/A/O/EI regime the abundance of PAOs and the efficiency of phosphorus removal can be significantly improved. This was experimentally demonstrated via a long-term test in two reactors operated as the new O/A/O/EI regime and the conventional A/O/A/O regime, respectively. The abundance of PAOs cultured in the O/A/O/EI reactor was about 11% higher than those in the conventional A/O/A/O reactor, which led to 15% of improved phosphorus removal efficiency. Moreover, this wastewater treatment regime based strategy did not decrease but slightly increase the nitrogen removal performance. Considering the huge quantities of wastewater treated daily, this strategy has a significant consequence from an ecological perspective.

Compared with other strategies such as pH adjustment and temperature control [6,15], this wastewater treatment regime based strategy does not require consumption of any additional chemicals and energy, which makes this strategy more economical and practical. This strategy can also integrate with the step-feeding mode easily, a practically effective method for minimizing the inhibitory effects of FNA, to gain a better nutrient removal performance. It was reported that step-feeding modes could greatly reduce the FNA inhibition influence as com- pared to dump-feeding [28]. By modifying the influent mode of O/A/O/EI regime, this wastewater treatment regime based strategy can easily combine with the feeding based strategy, which may cause further reduction of FNA inhibitory. Therefore, the strategy presented here might provide a practically promising solution to the "nitrogen-phosphorus challenge" faced by WWTPs. Furthermore, the enrichment of PAOs in the O/A/O/EI reactor is driven by the O/EI regime. It was reported that the O/EI regime could achieve very good phosphorus removal readily and steady when using glucose, a substrate usually considered being detrimental for PAO proliferations, as the sole carbon source

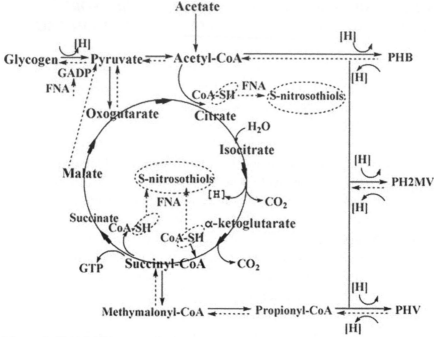

Figure 5: FNA inhibitory mechanisms on PAOs (dark solid line: anaerobic inhibitory mechanisms adapted from the literature [6]; dark dash line: oxic inhibitory mechanisms adapted from the literatures [4,5]).

[30]. Thus, the O/A/O/EI process reported in this paper may provide an ideal technology for BNR removal from carbohydrate-rich wastewaters. Generally, glucose or other carbohydrate compounds in domestic wastewater are at low levels, because it can be readily biofermented to volatile fatty acids in sewer systems. However, in some WWTPs where industrial or agricultural factories discharging carbohydrate-rich wastewaters are located nearby, or in some specific areas where the distance between the wastewater discharge sources and wastewater treatment unit is short (e.g., the highway rest areas, one of our parallel researches), wastewater carbohydrate may maintain at high levels. In these areas, the O/A/O/EI process may have an excellent application perspective, and the batch-scale study presented here may provide a useful reference for designs in future.

It should be noted that although the hydraulic retention time (HRT) between the O/A/O/EI and A/O/A/O reactors operated in this studies was maintained the same, the HRT controlled in the O/A/O/EI regime may slightly higher than that in the conventional BNR regime, because one cycle of the conventional BNR systems can be shortened to be 6 h via process optimization whereas the O/A/O/EI regime needs a relatively long idle period to enrich PAOs (e.g., 210 min). This characteristic implies that the proposed O/A/O/EI regime will increase the volumes of bioreactors when treating the same amount of wastewater. However, this drawback can be settled via reactor reconfiguration as proposed in Figure S1. Despite that this new strategy was demonstrated using sequencing batch reactors in this study (due to the availability of the equipment), it has also the potential to be applied in a continuous system. For a continuous-flow activated sludge system, an extra reactor for regurgitant sludge rest (3.5 h of the retention time seems to be enough) is required to set up in the side-stream for the enrichment of PAOs, and the construction invest of extra side-stream reactor is low, as compared with other strategies. It should also be emphasized that fullscale tests are required to fully evaluate the feasibility and potential of this strategy though excellent results have already obtained in our laboratory experiments.

8.4 METHODS

8.4.1 SYNTHETIC WASTEWATER

Synthetic wastewater used throughout these investigations, unless otherwise described, was the same and prepared daily. Acetate was used for the sole carbon source since it was the most common volatile fatty acids present in real domestic wastewaters [29]. KH_2PO_4 was selected as the phosphorus source. The chemical oxygen demand (COD) and ortho phosphate (PO_4^{3-}-P) concentrations in the wastewater were approximately maintained at 300 and 15 mg/L, respectively. Hence, the ratio of COD: PO_4^{3-}-P in the influent was controlled at 20 mg COD/(mg PO_4^{3-}-P), which was considered as being favorable to the growth of PAOs [2]. The concentrations of the other nutrients in the synthetic wastewater were the same and indicated below (per liter): 133.8 mg NH_4Cl, 0.5 mg $CaCl_2$, 0.5 mg $MgSO_4$, and 1 mL of a trace metals solution. The trace metals solution had been described in our previous publication [30].

8.4.2 OPERATION OF THE NEW AND CONVENTIONAL FOUR-STEP BNR PROCESSES

This study was conducted in two identical sequencing batch reactors with a working volume of 12 L each. Both reactors were seeded with activated sludge obtained from a WWTP in Changsha, PR China, which was operated as A^2/O process. The initial concentration of mixed liquor suspended solids (MLSS) was 3800 mg/L and mixed liquor volatile suspended solid (MLVSS) was 2400 mg/L. The activated sludge was maintained at 20±0.5 °C in a temperature controlled room. One reactor was performed as the developed O/A/O/EI regime while the other was operated as the classical four-step A/O/A/O regime in parallel. Both reactors were operated with three 8-h cycles daily, and each 8-h cycle of the O/A/O/EI regime consisted of a 120 min oxic phase, a 90 min anoxic phase, a 30 min oxic phase, a 30 min settling and decanting phase, and a 210 min idle phase. As comparison, the conventional four-step BNR regime was also operated according to the literature with minor revision [31,32], and each cycle of this regime contained a 90 min anaerobic period, a 120 min aerobic period, a 90 min anoxic period, a 30 min aerobic period, a 30 min settling/decanting period, and a 120 min idle period. For each cycle, certain volume supernatant was discharged from both reactors after the settling phase and was replaced with synthetic wastewater during the initial 5 min of first oxic phase (the O/A/O/EI reactor) and anaerobic phase (the A/O/A/O reactor), respectively. The HRT and sludge retention time (SRT) in the two reactors were controlled at approximately 16 h and 20 d, respectively. During anaerobic phase, the A/O/A/O reactor was mixed with a mechanical stirrer (150 rpm) while during the aerobic phase, air was supplied into both reactors at a flow rate of 15 L/min. The initial pH level in both reactors was controlled at 8.0 by adding 0.5 M HCl or 0.5 M NaOH solutions.

It should be noted that during the idle phase mixture stirring was not conducted in the routine operation of both reactors, but when cyclic tests were carried out, both reactors were mixed with a mechanical stirrer (150 rpm) to facilitate sampling. The mixed liquor samples were taken every 30 min and immediately filtered through a Whatmann GF/C glass microfiber filter (1.2 mm). The sludge sample was used to assay for MLSS, MLVSS, PHAs, and glycogen. The filtrate was used for the analyses of SOP, COD, NH_4^+-N, and NO_x^--N.

8.4.3 PHOSPHORUS REMOVAL TEST VIA CHEMICAL PRECIPITATION AT DIFFERENT PH

Phosphorus can be removed via chemical precipitation when some metal ions such as Ca^{2+}, Mg^{2+} are present in wastewater. With the increase of pH the chemical phosphorus precipitation was enhanced [22]. Hence, one batch test was performed without the activated sludge microorganisms to assess the effect of pH on chemical phosphorus precipitation. Firstly, 15 L synthetic wastewater mentioned above was divided evenly into 5 identical reactors with working volumes of 3.0 L each. Then, the reactors were added 0.5 M HCl or 0.5 M NaOH solutions to keep the pH value 6.5, 7.0, 7.5, 8.0, and 8.5, respectively, the other operational conditions were the same as the O/A/O/EI reactor described above expect that there was no sludge microorganisms. Finally, SOP in the supernatant of the 5 reactors was detected after several cycles. Therefore, it is readily to assess the effect of pH on chemical phosphorus precipitation via measuring the SOP concentration in supernatant.

Comparison of the effect of different FNA levels on PAO metabolisms between the two reactors. The A/OA/O reactor is developed from the conventional A/O phosphorus removal regime whereas the O/A/O/EI reactor is developed from the recently exploited O/EI regime, thus it is necessary to investigate whether there are different effects on PAO metabolisms between the two reactors even under the same FNA levels. The following batch experiment was executed to provide such support. Two identical sludge mixtures (2.4 L each) were respectively withdrawn from a WWTP in Changsha, PR China. The mixtures were centrifuged (5000 rpm for 5 min) and washed three times with tap water to remove the residual NH_4^+-N, NO_x^-- N, SOP, and COD. Then, they were resuspended in tap water with a final volume of 1.2 L each before being evenly divided into six reactors. The two groups of reactors (six each) were respectively operated as the same as the A/OA/O and O/A/O/EI reactors expect for the following differences. Allyl-Nthiourea (a nitrification inhibitor) was added at a concentration of 2 mg/L to each reactor to inhibit the nitrification according to the literature [33]. Temperature was controlled at 20±0.5 °C and pH was on-line controlled consistently at pre-designed set-point (pH=8.0±0.1) by a programmable logic controller using 0.5 M HCl solution and 0.5 M NaOH solution. Thus, the FNA concentrations of six reactors for each group were respectively controlled at 0, 0.05×10^{-3}, 0.15×10^{-3}, 0.26×10^{-3}, 0.38 3 1023, and 0.51×10^{-3} mgHNO$_2$-N/L through controlling nitrite concentration, pH, and temperature. It was reported that the FNA concentration could be calculated by the formula $S_{N-NO2}/(K_a \times 10^{pH})$ with K_a value determined by the formula $K_a = e^{(-2300/(T+273))}$ for a given temperature T (°C)24. When effluent SOP concentration among these reactors reached stable, cyclic studies were performed and the data were reported.

8.4.4 CHEMICAL AND MICROBIAL ANALYSES

COD, SOP, nitrite, nitrate, ammonia, MLSS, and MLVSS were measured by standard methods [34]. The determinations of glycogen, poly-3-hydroxybutyrate (PHB), poly-3-hydroxyvalerate (PHV), and poly-3-hydroxy-2-methylvalerate (PH2MV) were measured according to our previous publication [30]. The PHAs were the summation of PHB, PHV, and PH2MV.

The fluorescence in situ hybridization (FISH) with 16s rRNA-targeted oligonucleotide probes was carried out to quantify the abundances of PAOs and GAOs, and the methods were the same as described in the literature [17]. Briefly, sludge samples were taken and fixed in 4% formaldehyde for 20 h at 4 °C and then subjected to freeze-thaw treatment in order to enhance the penetration of oligonucleotide probes. Cell samples were attached to poly-L-lysine coated slides and dehydrated with ethanol. The following hybridization and washing procedures were the same as that in the literature [35]. For quantitative analysis, 20 microscopic fields were analyzed for the hybridization of individual probes using a confocal scanning laser microscope (FV 500) with image database software (VideoTesT Album3.0). The oligonucleotide probes specific for PAOs, GAOs, and total bacteria, which were respectively labeled with 5'AMCA, 5'Cy3, and 5'FITC, were listed in Table S2.

REFERENCES

1. Mino, T., Van Loosdrecht, M. C. M. & Heijnen, J. J. Microbiology and biochemistry of the enhanced biological phosphate removal process. Water Res. 32, 3193–3207 (1998).
2. Oehmen, A. et al. Advances in enhanced biological phosphorus removal: from micro to macro scale. Water Res. 41, 2271–2300 (2007).
3. Smolders, G. J. F., Van der Meij, J., Van Loosdrecht, M. C. M. & Heijnen, J. J. Model of the anaerobic metabolism of the biological phosphorus removal process: stoichiometry and pH influence. Biotechnol. Bioeng. 43, 461–470 (1994).
4. Martín, H. G. et al. Metagenomic analysis of two enhanced biological phosphorus removal (EBPR) sludge communities. Nat Biotechnol. 24, 1263–1269 (2006).
5. Pijuan, M., Ye, L. & Yuan, Z. Free nitrous acid inhibition on the aerobic metabolism of poly-phosphate accumulating organisms. Water Res. 44, 6063–6072 (2010).
6   Zhou, Y., Oehmen, A., Lim, M., Vadivelu, V. & Ng, W. J. The role of nitrite and free nitrous acid (FNA) in wastewater treatment plants. Water Res. 45, 4672–4682 (2011).
7. Zhou, Y., Ganda, L., Lim, M., Yuan, Z. & Ng, W. J. Response of poly-phosphate accumulating organisms to free nitrous acid inhibition under anoxic and aerobic conditions. Bioresour. Technol. 116, 340–347 (2012).
8. Ye, L., Pijuan, M. & Yuan, Z. The effect of free nitrous acid on key anaerobic processes in enhanced biological phosphorus removal systems. Bioresour. Technol. 130, 382–389 (2013).
9. Zeng, W., Yang, Y., Li, L., Wang, X. & Peng, Y. Effect of nitrite from nitritation on biological phosphorus removal in a sequencing batch reactor treating domestic wastewater. Bioresour. Technol. 102, 6657–6664 (2011).
10. Pijuan, M. & Yuan, Z. Development and optimization of a sequencing batch reactor for nitrogen and phosphorus removal from abattoir wastewater to meet irrigation standards. Water Sci. Technol. 61, 2105–2112 (2010).
11. Zhou, Y., Pijuan, M. & Yuan, Z. Free nitrous acid inhibition on anoxic phosphorus uptake and denitrification by poly-phosphate accumulating organisms. Biotechnol. Bioeng. 98, 903–912 (2007).
12. Zhou, Y. et al. Free nitrous acid (FNA) inhibition on denitrifying poly-phosphate accumulating organisms (DPAOs). Appl. Microbiol. Biotechnol. 88, 359–369 (2010).

13. Saito, T., Brdjanovic, D. & van Loosdrecht, M. C. M. Effect of nitrite on phosphate uptake by phosphate accumulating organisms. Water Res. 38, 3760–3768 (2004).
14. Zhou, Y. A. N., Pijuan, M., Zeng, R. J. & Yuan, Z. Free nitrous acid inhibition on nitrous oxide reduction by a denitrifying-enhanced biological phosphorus removal sludge. Environ. Sci. Technol. 42, 8260–8265 (2008).
15. Zhou, Y., Pijuan, M. & Yuan, Z. Development of a 2-sludge, 3-stage system for nitrogen and phosphorous removal from nutrient-rich wastewater using granular sludge and biofilms. Water Res. 42, 3207–3217 (2008).
16. Wang, D. et al. Improved biological phosphorus removal performance driven by the aerobic/extended-idle regime with propionate as the sole carbon source. Water Res. 46, 3868–3878 (2012).
17. Wang, D. et al. Inducing mechanism of biological phosphorus removal driven by the aerobic/extended-idle regime. Biotechnol. Bioeng. 109, 2798–2807 (2012).
18. Wang, D. et al. Evaluation of the feasibility of alcohols serving as external carbon sources for biological phosphorus removal induced by the oxic/extended-idle regime. Biotechnol. Bioeng. 110, 827–837 (2013).
19. Wang, D. et al. Effect of initial pH control on biological phosphorus removal induced by the aerobic/extended-idle regime. Chemosphere 90, 2279–2287 (2013).
20. Marcelino, M., Guisasola, A. & Baeza, J. A. Experimental assessment and modelling of the proton production linked to phosphorus release and uptake in EBPR systems. Water Res. 43, 2431–2440 (2009).
21. Janssen, P. M. J., Meinema, K. & van der Roest, H. F. Biological Phosphorus Removal-Manual for Design and Operation. IWA Publishing, London (2002).
22. Maurer, M., Abramovich, D., Siegrist, H. & Gujer, W. Kinetics of biologically induced phosphorus precipitation in wastewater treatment. Water Res. 33, 484–493 (1999).
23. Oehmen, A., Zeng, R. J., Yuan, Z. G. & Keller, J. Anaerobic metabolism of propionate by polyphosphate-accumulating organisms in enhanced biological phosphorus removal systems. Biotechnol. Bioeng. 91, 43–53 (2005).
24. Anthonisen, A. C., Loehr, R. C., Prakasam, T. B. S. & Shinath, E. G. Inhibition of nitrification by ammonia and nitrous acid. J Water Pollut Control Fed. 48, 835–852 (1976).
25. Coats, E. R., Mockos, A. & Loge, F. J. Post-anoxic denitrification driven by PHA and glycogen within enhanced biological phosphorus removal. Bioresour. Technol. 102, 1019–1027 (2011).
26. Xu, X., Liu, G. & Zhu, L. Enhanced denitrifying phosphorous removal in a novel anaerobic/aerobic/anoxic (AOA) process with the diversion of internal carbon source. Bioresour. Technol. 102, 10340–10345 (2011).
27. Tay , C., Garlapati, V. K., Guisasola, A. & Baeza, J. A. The selective role of nitrite in the PAO/GAO competition. Chemosphere 93, 612–618 (2013).
28. Vargas, M., Guisasola, A., Artigues, A., Casas, C. & Baeza, J. A. Comparison of a nitrite-based anaerobic-anoxic EBPR system with propionate or acetate as electron donors. Process Biochem. 46, 714–720 (2011).
29. Chen, Y., Randall, A. A. & McCue, T. The efficiency of enhanced biological phosphorus removal from real wastewater affected by different ratios of acetic to propionic acid. Water Res. 38, 27–36 (2004).
30. Wang, D. B. et al. The probable metabolic relation between phosphate uptake and energy storages formations under single-stage oxic condition. Bioresour. Technol.

100, 4005–4011 (2009).Pai, T. Y. et al. Microbial kinetic analysis of three different types of EBNR process.Chemosphere 55, 109–118 (2004).

31. Uygur, A. Specific nutrient removal rates in saline wastewater treatment using sequencing batch reactor. Process Biochem. 41, 61–66 (2006).

32. Chen, Y., Wang, D., Zhu, X., Zheng, X. & Feng, L. Long-Term Effects of copper nanoparticles on wastewater biological nutrient removal and N_2O generation in the activated sludge process. Environ. Sci. Technol. 46, 12452–12458 (2012).

33. APHA. Standard Methods for the Examination of Water and Wastewater. 20 edn, American Public Health Association (1998).

34. Nielsen, A. T. et al. A. Identification of a novel group of bacteria in sludge from a deteriorated biological phosphorus removal reactor. Appl. Environ. Microbiol. 65, 1251–1258 (1999).

There are several supplemental files that are not available in this version of the article. To view this additional information, please use the citation on the first page of this chapter.

PART IV

XENOBIOTICS

CHAPTER 9

Bacterial Consortium and Axenic Cultures Isolated from Activated Sewage Sludge for Biodegradation of Imidazolium-based Ionic Liquid

MARTA MARKIEWICZ, JOANNA HENKE,
ANNA BRILLOWSKA-DABROWSKA, STEFAN STOLTE,
JUSTYNA LUCZAK, and CHRISTIAN JUNGNICKEL

9.1 INTRODUCTION

Ionic liquids (ILs) are chemicals usually composed of large asymmetric, organic cation and organic or inorganic anion. Physical and chemical properties of this group of compounds can vary significantly what allows them to be designed for a particular purpose (Krossing et al. 2006). The last decade has shown a growing interest in the application of ILs in gas storage and separation, catalysis, electrodeposition of metals, waste and biomass reprocessing, energy production, etc. (Kragl et al. 2002; Jiang et al. 2006; Plechkova and Seddon 2008). When applied in such industrial

Bacterial Consortium and Axenic Cultures Isolated from Activated Sewage Sludge for Biodegradation of Imidazolium-based Ionic Liquid © *Markiewicz M, Henke J, Brillowska-Dabrowska A, Stotle S, Luczak J, and Jungnickel C.* International Journal of Environmental Science and Technology *11,7 (2014), doi: 10.1007/s13762-013-0390-1. Licensed under a Creative Commons Attribution 2.0 Generic License, http://creativecommons.org/licenses/by/2.0/.*

processes, ILs will inevitably emerge in wastewaters and might end up in natural soils or water bodies by breaking through treatment systems or due to the accidental release during transport and storage. Although, the low volatility of ILs can be an advantage in reducing air emissions and thereby decreasing the risk of human exposure, the relatively high toxicity and resistance to biotic and abiotic degradation that could be observed for some of the ILs structures is a concern (Romero et al. 2008).

Biodegradation of substituted imidazolium cation was examined in detail by a number of research groups (Gathergood et al. 2006; Romero et al. 2008; Stolte et al. 2008; Markiewicz et al. 2009; Abrusci et al. 2010; Coleman and Gathergood 2010). Stolte et al. conducted a comprehensive study of biodegradation of 1-methyl-3-alkylimidazolium chlorides. No primary biodegradation was observed for methyl- to butyl-substituted compounds, even those containing oxygen atoms introduced into the alkyl chain (as ethers or terminal hydroxyl groups) which are known to increase biodegradability. Better biodegradability was observed for higher homologs with 1-methyl-3-octy-limidazolium chloride reaching 100% biodegradation after 24 day of the test (Stolte et al. 2008). In most IL biodegradation tests, activated sewage sludge is used as an inoculum as it is composed of multiple species with a wide taxonomic diversity and is therefore more likely to contain specie or species capable of degrading ILs (Gathergood et al. 2006; Romero et al. 2008; Stolte et al. 2008; Markiewicz et al. 2009). In the previously mentioned work, Stolte et al. examined the ability of commercially available freeze-dried bacteria mixture to biodegrade ILs and concluded that it was unable to degrade any of tested compounds. Modelli et al. used inoculum derived from top soil for biodegradation of ILs under ASTM D 5988-96 test conditions and found it capable of degrading 1-butyl-3-methylimidazolium chloride (Modelli et al. 2008). Abrusci et al. used a strain of *Sphingomonas paucimobilis* isolated from cinematographic film for biodegradation of some common ILs and found it predisposed to degrade many tested compounds including those previously reported to be non-biodegradable. However, a somehow surprising trend was observed by Abrusci et al. for imidazolium chlorides showing highest biodegradation for ethyl- and lowest for octyl- and decyl-substituted compounds (Abrusci et al. 2010).

Some bacterial species are capable of tolerating or even growing in the presence of xenobiotics (Isken and deBont 1998; Takenaka et al. 2007). In pursuit of understanding the mechanisms of bacterial resistance and adaptation to xenobiotics, it was revealed that resistance is either a natural property of a species or is acquired by genetic changes (Weber and deBont 1996; Pham et al. 2009). The ability to degrade xenobiotics is not simply a function of the amount of biomass of inoculum, although very often a correlation between the two exists. Activated sewage sludge community is often used in biodegradation tests as it is expected that degraders of xenobiotics will be encountered among the multitude of species. Assuring this by adding specialized species is known as (bio)augmentation and was previously suggested as a promising strategy to enhance the degradative capacity of soils and sewage sludge (Limbergen et al. 1998; Pieper and Reineke 2000). For this reason, an attempt to identify the exact microbial strains which might be partaking in the process of biological degradation of ILs was undertaken here in order to uncover specie or species especially predisposed to degrade ILs. Would the attempt be successful, it might present a very useful tool in enhancing biodegradation of ILs by augmenting indigenous microbial communities. As activated sewage sludge was clearly proven to be the most potent in biodegrading ILs, it was chosen as a starting point. For the current test, 1-methyl-3-octylimidazolium chloride was selected since it was previously shown to be biodegradable in activated sewage sludge, and therefore, it will allow for comparative statements to be drawn. A nearly 30-fold increase in maximum biodegradable concentration and growing rates of degradation resulting from pre-exposition of activated sewage sludge community to this IL were reported before. Moreover, a break-down of the imidazolium ring was observed (Markiewicz et al. 2011). In the course of the current work, nine strains of bacteria from adapted activated sewage sludge were isolated, and the influence of 1-methyl-3-octylimidazolium chloride on their growth was tested. The performance of consortium of all isolated strains and of each strain individually was verified in 'Manometric respirometry' biodegradation test using the same IL. Results of these tests are presented hereby.

9.2 MATERIALS AND METHODS

9.2.1 CHEMICALS

The investigated ionic liquid 1-methyl-3-octylimidazolium chloride was purchased from Merck KGaA (Germany) with a purity of C98%. The molecular weight, chemical formula and molecular structure are shown in Table 1.

9.2.2 SELECTION OF RESISTANT ISOLATES

The experiments were carried out at the Department of Chemical Technology, Gdańsk University of Technology and at the Center for Environmental Research and Sustainable Technology, University of Bremen. The sewage was sampled in June 2009, and the measurements were carried out in the subsequent 10 months. Activated sewage sludge was obtained from aeration tank of municipal wastewater treatment plant (WWTP) 'Wschód' in Gdańsk, Poland, and used for biodegradation experiment during which it was subjected to selective pressure of [OMIM][Cl] as described in Markiewicz et al. (2009). Subsequently, microbial strains were isolated and identified as described below.

9.2.3 BACTERIAL SPECIES IDENTIFICATION

Activated sewage sludge, obtained from the last phase of biodegradation test, was diluted with saline (0.85% NaCl) solution. Samples diluted from 1 to 5 times were then inoculated into Petri dishes containing Luria Agar (LA) using spread-plate method and incubated in 16 °C for 48 h. To obtain pure cultures, single colonies were chosen and inoculated to Petri dishes with LA agar, using streak plate method and incubated. Single colonies

were inoculated to test tubes with 2 mL of Luria Broth (LB) and cultivated. After 48 h, 100 lL of each liquid culture was inoculated into Petri
dishes containing LA agar by spread-plate method and cultured for 48 h at
16 °C. Bacterial isolates from activated sewage sludge were subjected to
Gram staining and observed under optical microscope.

9.2.4 MOLECULAR IDENTIFICATION OF BACTERIAL ISOLATES

DNA was purified by mean of Genomic Mini Kit (A&A Biotechnology,
Poland). PCR with universal 16S rDNA primers: 16S-for 5' GGACTAC-
CAGGGTATCTAATC 3' and 16S-rev 5' GATCCTGGCTCAGGATGAAC
3' and REDTaq® ReadyMix PCR Reaction Mix (Sigma-Aldrich®) were
performed with isolated DNA of nine bacterial strains in a volume of 20
lL. The time–temperature profile for PCR was 35 cycles of 30 s at 94 °C,
30 s at 55 °C and 45 s at 72 °C, preceded by initial denaturation for 10 min
at 95 °C. The presence of specific PCR products of approximately 800 bp
was examined using electrophoresis on a 1% agarose gel and staining with
ethidium bromide.

The PCR products were purified with High Pure PCR Product Purification Kit (Roche) and sequenced (Genomed, Poland). The identification
of isolates was performed by comparison of obtained sequences against
the sequences from GenBank (http://blast.ncbi.nlm.nih.gov)

9.2.5 GROWTH INHIBITION TEST

Fresh bacterial cultures of the nine isolates were grown in LB broth (cell
density $4 \cdot 10^8$ cells/L). IL ([OMIM][Cl]) was used in four concentrations:
0.2; 2; 20 and 200 mM. Positive control samples, containing LB, inoculum
and sodium glutamate with the same carbon content as respective samples
with IL—one for each microorganism—were prepared. Additionally, the
sample with LB and IL only served as a negative control. All prepared
samples were then incubated in 16 °C for 48 h. After incubation, in order
to determine the growth rate in each sample, 200 μL of each solution was
transferred to the disposable microplate, and the quantity of bacterial cells

was determined spectropho- tometrically using a multilabel plate reader Wallac 1420 VICTOR³-V (λ = 595 nm wavelength).

9.2.6 BIODEGRADATION TESTS

Ready biodegradability tests of 1-methyl-3-octylimidazolium chloride in concentration of 0.2 mM using 1 mL of axenic cultures (resulting in cell density of $8 \cdot 10^4$ cells/L in each culture) of nine isolates were performed according to OECD 301 F 'Manometric respirometry' procedure (OECD 1992). In this test, biodegradation is measured as a decrease in pressure in gas-tight test vessel caused by depletion of oxygen used for aerobic degradation of IL reduced by the blank sample value (sample showing only endogenous respiration of bacteria, without addition of test compound) with respect to the theoretical amount of oxygen necessary to completely oxidize the compound tested. Since almost no biodegradation was observed, the same test was repeated for a consortium composed of all nine isolates (cell density $8 \cdot 10^5$ cells/L of all strains in total) in a concentration of IL previously reported to be low enough for biodegradation with activated sewage sludge to occur (0.25 mM) (Stolte et al. 2008). Samples containing activated sewage sludge (cell density $10 \cdot 10^7$ cells/L) derived from 'Wschód' in Gdańsk, were also employed in the test for the sake of comparison.

9.3 RESULTS AND DISCUSSION

9.3.1 ISOLATION AND IDENTIFICATION OF SEWAGE SLUDGE BACTERIA

Nine different isolates of microorganisms were cultivated from original biodegradation test utilizing activated sewage sludge. After DNA isolation, followed by 16S rDNA PCR and product purification, the DNA sequencing was conducted. The obtained sequences were compared with GenBank data (NCBI 2011), and microbial species were identified.

Table 2 summarizes the results of identification and Gram staining. Figure 1 presents the phylogenetic tree of identified species. The sequence alignment and bootstrap values were calculated using CLUSTALW2.012.

9.3.2 GROWTH INHIBITION

The inhibition of growth of nine isolated bacterial strains by 1-methyl-3-octylimidazolium chloride [OMIM][Cl] in four concentrations covering four orders of magnitude was examined. Therefore, the strains were incubated in the presence of IL for 48 h, and afterward, the cell density was determined spectrophotometrically. The results were expressed as a percent of growth in relation to a positive control with sodium glutamate (Fig. 2). Concentration-dependent decrease in growth was found. In 0.2 mM [OMIM][Cl], most of strains showed a similar decrease in growth (10–20%) except strains 4 and 5 where growth reached approximately 150 and 120% of positive control, respectively. This might suggest that those two species are especially predisposed to utilizing [OMIM][Cl] as car-

Table 2: Microbial isolates identified by sequencing of 16S rDNA sequences comparison with GenBank data (NCBI 2011)

No.	Organism	Gram
1	*Flavobacterium* sp. WB3.2-27	–
2	*Shewanella putrefaciens* CN-32	–
3	*Moraxellaceae Bacterium* MAG	–
4	*Flavobacterium* sp. FB7	–
5	*Microbacterium keratanolyticum* AO17b	+
6	*Flavobacterium* sp. WB 4.4-116	–
7	*Arthrobacter* sp. SPC 26	+
8	*Rhodococcus* sp. PN8	+
9	*Arthrobacter protophormiae* strain DSM 20168	+

bon source or, more probably, might have been a result of hormesis. It is believed that when exposed to low doses of toxin, organisms exhibit increased growth by investing larger amounts of energy in reproduction in order to assure survival (Calabrese 2005). It is possible that for *Flavobacterium* sp. FB5 and *Microbacterium keratanolyticum*, this concentration of IL triggered the hormetic effect.

Increasing IL concentration to 2 mM caused 50–60% inhibition of growth for most of strains. *Microbacterium keratanolyticum* proved to be particularly sensitive (75% inhibition) supporting hypothesis of hormesis. No inhibition of growth of *Arthrobacter* sp. SPC 26 and only minor inhibition of *Rhodococcus* sp. PN8 were found. Enhanced growth of *Arthrobacter* sp. as compared 2 mM sample can again be explained by hormesis. *Rhodococcus* exhibited steady growth in both 0.2 and 2 mM samples, making this strain the most tolerant for IL.

Most of isolated strains proved to be slightly more resistant to IL than previously reported by Łuczak et al. (2010), where at least 80% growth inhibition was observed in 4 mM [OMIM][Cl] for Gram-negative *Escherichia coli* and in 2 mM [OMIM][Cl] for three other Gram-positive bacterial strains. The growth observed for samples containing 20 mM IL solution was almost negligible and completely inhibited in 200 mM solutions (results not shown).

A variety of negative effects of chemicals on function of bacterial cells are known. For example, membrane permeabilization is caused by adsorption of molecules, resulting in leakage of cytoplasm and micromolecules or simply inability to maintain cell integrity. Other effects include the decrease in the cell energy status caused by passive flux of ions through the membrane, the disturbance of functions of other (not involved in energy transduction) proteins present in the membrane as well as the distortions in fluidity and the hydration of the membrane surface (Isken and deBont 1998). Bacterial cells might counteract these effects by adaptation at the level of cytoplasmic membrane, including changes in level of saturation of fatty acids which changes the fluidity of membrane in order to compensate the negative effect of chemicals (known as homeoviscosic adaptation Heipieper and deBont 1994); reduction in cell hydrophobicity by altering the content of lipopolysaccharide (LPS) and modification of porines which was proven to increase the solvent tolerance of microorganism; solvents

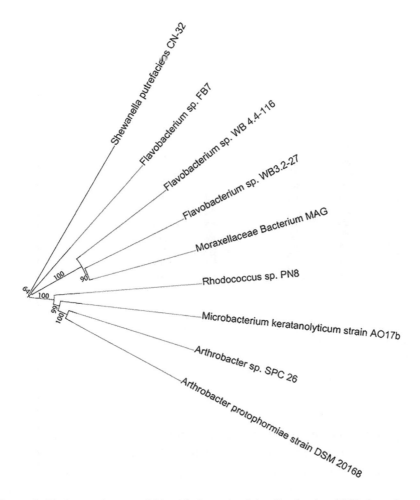

Figure 1: Phylogenetic tree of identified species (visualized using iTOL Letunic and Bork 2007)

degradation into less- or nontoxic substances and the solvent active excretion from the cell (Isken and deBont 1998). It would be expected that bacteria able to survive in unfavourable conditions created by the presence of ionic liquids would employ one of these mechanisms in order to survive, yet not necessarily will be able to break them down. The difference in cell wall construction of Gram-positive and Gram-negative bacteria might be responsible for their slightly different resistance toward chemicals. Even

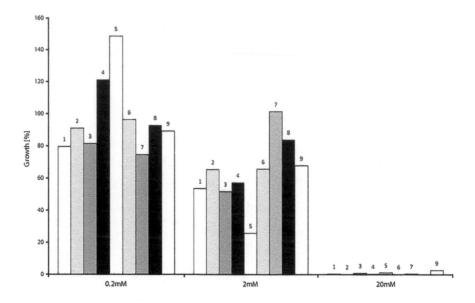

Figure 2: Growth of the selected bacterial strains in solution containing various concentrations of [OMIM][Cl] is displayed as a percent of positive control. Strains were numbered in accordance to Table 2

though Gram-negative cell wall is much thinner, it is covered with an additional lipid membrane acting as a barrier for many biocides, and thus, Gram-negative microorganisms show lower sensitivity to organic chemials including surfactants (Blazevic 1976; Volkering et al. 1995). The same regularity, though only slightly pronounced, was observed by Łuczak et al. for ILs (Luczak et al. 2010). Nevertheless, no obvious differences in growth inhibition between Gram-positive and Gram-negative bacteria were observed within this work.

9.3.3 BIODEGRADATION TESTS

All isolated stains from previous experiment (Markiewicz et al. 2009) were tested in biodegradation tests according to OECD 301 F (Manometric respirometry) in order to determine the mineralization of [OMIM][Cl]. None of the isolates exhibited any significant levels of biodegradation of 1 methyl-3-octylimidazolium chloride when applied as an axenic culture. A maximum value of 8% was observed for *Microbacterium keratanolyticum* (results not shown) which was expected as this strain showed the highest growth rate in growth inhibition test in 0.2 mM sample. It is probable that not an axenic culture but a consortium of two or more strains is necessary to conduct full biodegradation of IL. Numerous examples of such situations can be found in literature. Consortia of bacteria were found to degrade azo-dyes, crude oil hydrocarbons, atrazine, etc., more efficiently than individual strains (Zwieten and Kennedy 1995; Khehra et al. 2005; Adebusoye et al. 2007). There are a number of interactions within a consortium that lead to degradation of a xenobiotic and that cannot occur in axenic culture. These are unfortunately very difficult to uncover. It may be that the first step is conducted by one organism resulting in an intermediate that is then transformed further by another organism, and this cascade proceeds to full mineralization. At some point, more than one organism can be involved in the degradation of intermediates leading to different products and consequent ramification of the metabolic pathway. It is also possible that some members of the consortia do not take part in the degradation of the xenobiotic, but support the primary degraders providing them with essential nutrients, e.g., vitamins, amino acids or creating appropriate environmental conditions such as removing toxins and adjusting oxygen levels. It is therefore far more likely that enhanced degradation can be achieved by augmentation with a consortium rather that an axenic culture (Grady 1985; Schink 2002).

The results of the biodegradation test with the consortium formed by mixing all nine isolates are shown in Fig. 3a. Biodegradation of sodium glutamate was also conducted to examine the viability of the inoculum. Additionally, biodegradation of both [OMIM][Cl] and sodium glutamate conducted by activated sewage sludge is displayed for comparison.

It is clear that both microbial communities were viable as sodium glutamate was completely degraded in both cases—degradation in activated sewage sludge was completed within 4 days and in the mixed strain consortium in 16 days. [OMIM][Cl] was degraded by activated sewage sludge at almost 60% by the end of the test which corresponds theoretically to a degradation of the octyl side chain without degradation of the imidazolium ring. Nevertheless, [OMIM][Cl] could not be classified as readily biodegradable on the basis of this result. Biodegradation of [OMIM][Cl] by the mixed strain consortium occurred slower and reached slightly above 30% on the twenty-eighth day of the test, most probably this also involved the biodegradation of the side chain. The differences in the degradation rates are possibly due to the different cell densities in the consortium and activated sewage sludge as well as different microbial composition. The dry mass cannot be used as a direct comparison, as the dry mass of sewage is partially comprised of a mineral fraction, extracellular polymeric substance as well as other organisms like fungi and protozoa. Therefore, to obtain a comparison of the various cultures, the cell density was determined by counting the colony-forming units. When normalized for cell density (Fig. 3b), the rate of biodegradation is much higher in case of mixed strain consortium.

This difference in rates is of course only an indicative result, since the cell density of sewage is several orders of magnitude higher (and therefore, the % degradation/CFU will always be lower). However, it highlights the potential that such mixed strain consortia might have.

9.4 CONCLUSION

Docherty et al. previously reported on changes in microbial genetic patterns of organisms cultivated in media containing ILs, suggesting that enrichment of certain species occurred (Docherty et al. 2007). We decided to follow this lead and isolated bacterial strains from activated sewage sludge subjected to selective pressure of 1-methyl-3-octy- limidazolium chloride.

Nine bacterial isolates were identified. Three of them belonged to *Flavobacteriaceae* family, two to family *Micrococcaceae* and the remaining strains to the families: *Shewanellacae, Moraxellaceae, Microbacteriaceae, Nocardiaceae*. All strains grew well in 0.2 mM [OMIM][Cl] and were inhibited in 40% on average in 2 mM solutions. Any higher concentrations inhibited growth almost completely.

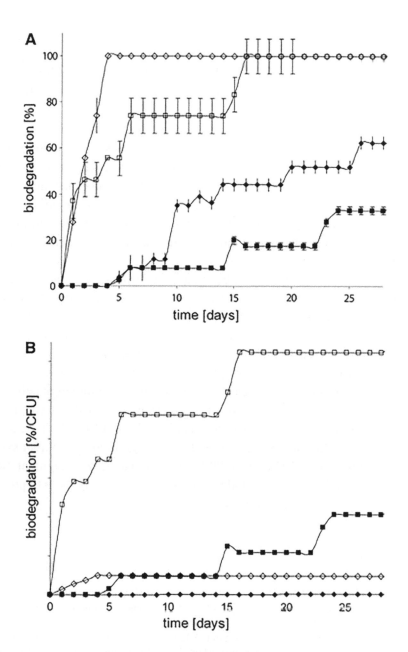

Figure 3: Biodegradation (a) and biodegradation normalized for cell density (b) of [OMIM] [Cl] (closed symbols) and sodium glutamate (open symbols) by activated sewage sludge (diamonds) and consortium of nine isolated strains (squares)

Axenic cultures of single isolates were proven to be rather inefficient in degrading [OMIM][Cl] and application of a consortium composed of all mixed strains resulted in 30% ultimate degradation. In the same conditions, activated sewage sludge organisms degraded almost 60% of the tested compound which corresponds to full degradation of alkyl substituents. The lower result obtained for the mixed strain consortium might have been a result of lower cell density in these samples. A higher degradation rate was observed in case of mixed strains when results were normalized for cell density in consortium. It is possible that not all of isolates involved in ILs metabolism were selected or that isolated microorganisms were a part of more complex consortium connected by symbiotic relations with other organisms and that, in their presence, even higher metabolic capacities could have been achieved.

We have recently shown that adapted microbial communities can degrade [OMIM][Cl] faster and can withstand higher concentrations of that IL without an inhibitory effect compared with non-adapted communities (Markiewicz et al. 2011). Therefore, more research is needed in order to uncover species involved in [OMIM][Cl] degradation and to examine the feasibility of using adapted single strains or consortia in enhancing degradative abilities of indigenous microorganisms. Obtaining high cell density inocula of strains isolated here and comparing them with activated sludge of the same cell density would unequivocally confirm their better performance. Additionally, simulation tests in high biomass content systems (e.g., OECD tests 303 OECD 2001) in combination with augmentation of freshly sampled activated sludge with nine isolates could help to prove the technological viability of this approach.

REFERENCES

1. Abrusci C, Palomar J, Pablos JL, Rodriguez F, Catalina F (2010) Efficient biodegradation of common ionic liquids by Sphingomonas paucimobilis bacterium. Green Chem 3:709–717
2. Adebusoye SA, Ilori MO, Amund OO, Teniola OD, Olatope SO (2007) Microbial degradation of petroleum hydrocarbons in a polluted tropical stream. World J Microbiol Biotechnol 23(8): 1149–1159
3. Blazevic DJ (1976) Current taxonomy and identification of nonfermentative gram negative bacilli. Hum Pathol 7(3):265–275 Calabrese EJ (2005) Paradigm lost, para-

digm found: the re-emergence of hormesis as a fundamental dose response model in the toxicological sciences. Environ Pollut 138:378–411

4. Coleman D, Gathergood N (2010) Biodegradation studies of ionic liquids. Chem Soc Rev 39:600–637
5. Docherty KM, Dixon JK, Jr CFK (2007) Biodegradability of imidazolium and pyridinium ionic liquids by an activated sludge microbial community. Biodegradation 18:481–493
6. Gathergood N, Scammells PJ, Garcia MT (2006) Biodegradable ionic liquids Part III: the first readily biodegradable ionic liquids. Green Chem 8:156–160
7. Grady LCPJ (1985) Biodegradation: its measurement and microbiological basis. Biotechnol Bioeng 27:660–674
8. Heipieper HJ, deBont JAM (1994) Adaptation of *Pseudomonas putida* S12 to ethanol and toluene at the level of fatty acid composition of membranes. Appl Environ Microbiol 60(12): 4440–4444
9. Isken S, deBont JAM (1998) Bacteria tolerant to organic solvents. Extremophiles 2:229–238
10. Jiang T, Brym MJC, Dube G, Lasia A, Brisard GM (2006) Electrodeposition of aluminium from ionic liquids: part II— studies on the electrodeposition of aluminum from aluminum chloride ($AICl_3$)—trimethylphenylammonium chloride (TMPAC) ionic liquids. Surf Coat Technol 201:10–18
11. Khehra MS, Saini HS, Sharma DK, Chadha BS, Chimni SS (2005) Comparative studies on potential of consortium and constituent pure bacterial isolates to decolorize azo dyes. Water Res 39(20):5135–5141
12. Kragl U, Eckstein M, Kaftzik N (2002) Enzyme catalysis in ionic liquids. Curr Opin Biotechnol 13:565 571
13. Krossing I, Slattery JM, Daguenet C, Dyson PJ, Oleinikova A, Weingaertner H (2006) Why are ionic liquids liquid? A simple explanation based on lattice and solvation energies. J Am Chem Soc 128:13427–13434
14. Letunic I, Bork P (2007) Interactive tree of life (iTOL): an online tool for phylogenetic tree display and annotation. Bioinformatics 23(1):127–128
15. Limbergen HV, Top EM, Verstraete W (1998) Bioaugmentation in activated sludge: current features and future perspectives. Appl Microbiol Biotechnol 50:16–23
16. Łuczak J, Jungnickel C, Łącka I, Stolte S, Hupka J (2010) Antimicrobial and surface activity of 1-alkyl-3-methylimidazolium derivatives. Green Chem 12:593–601
17. Markiewicz M, Jungnickel C, Markowska A, Szczepaniak U, Paszkiewicz M, Hupka J (2009) 1-Methyl-3-octylimidazolium chloride—sorption and primary biodegradation analysis in activated sewage sludge. Molecules 14:4396–4405
18. Markiewicz M, Stolte S, Lustig Z, Łuczak J, Skup M, Hupka J, Jungnickel C (2011) Influence of microbial adaption and supplementation of nutrients on the biodegradation of ionic liquids in sewage sludge treatment processes. J Hazard Mater 195(15):378–382
19. Modelli A, Sali A, Galletti P, Samori C (2008) Biodegradation of oxygenated and non-oxygenated imidazolium-based ionic liquids in soil. Chemosphere 73(8):1322–1327
20. NCBI (2011) GenBank. From http://www.ncbi.nlm.nih.gov/genbank/ OECD (1992)
21. OECD guideline for testing of chemicals 301—ready biodegradability
22. OECD (2001) Simulation test—aerobic sewage treatment

23. Pham TPT, Cho C-W, Jeon C-O, Chung Y-J, Lee M-W, Yun Y-S (2009) Identification of metabolites involved in the biodegradation of ionic liquid 1-butyl-3-methylpyridinium bromide by activated sludge microorganisms. Environ Sci Technol 43:516–521

24. Pieper DH, Reineke W (2000) Engineering bacteria for bioremediation. Curr Opin Biotechnol 11:262–270

25. Plechkova NV, Seddon KR (2008) Applications of ionic liquids in the chemical industry. Chem Soc Rev 37:123–150

26. Romero A, Santos A, Tojo J, Rodrıguez A (2008) Toxicity and biodegradability of imidazolium ionic liquids. J Hazard Mater 151:268–273

27. Schink B (2002) Synergistic interactions in the microbial world. Antonie Van Leeuwenhoek 81(1–4): 257–261

28. Stolte S, Abdulkarim S, Arning J, Blomeyer-Nienstedt A-K, Bottin- Weber U, Matzke M, Ranke J, Jastorff B, Thöming J (2008) Primary biodegradation of ionic liquid cations, identification of degradation products of 1-methyl-3-octyl-imidazolium chloride and electrochemical waste water treatment of poorly biodegradable compounds. Green Chem 10:214–224

29. Takenaka S, Tonoki T, Taira K, Murakami S, Aoki K (2007) Adaptation of Pseudomonas sp. strain 7–6 to quaternary ammonium compounds and their degradation via dual pathways. Appl Environ Microbiol 73(6):1797–1802

30. Volkering F, Breure AM, Andel JGv, Rulkens WH (1995) Influence of nonionic surfactants on bioavailability and biodegradation of polycyclic aromatic hydrocarbons. Appl Environ Microbiol 61(5):1699–1705

31. Weber FJ, deBont JAM (1996) Adaptation mechanisms of microor- ganisms to the toxic effects of organic solvents on membranes. Biochim Biophys Acta 1286:225–245

32. Zwieten LV, Kennedy IR (1995) Rapid degradation of atrazine by Rhodococcus sp. NI86/21 and by an atrazine-perfused soil. J Agric Food Chem 43(5):1377–1382

There are several supplemental files that are not available in this version of the article. To view this additional information, please use the citation on the first page of this chapter.

PART V

METHANE PRODUCTION

CHAPTER 10

Zero Valent Iron Significantly Enhances Methane Production from Waste Activated Sludge by Improving Biochemical Methane Potential Rather than Hydrolysis Rate

YIWEN LIU, QILIN WANG, YAOBIN ZHANG AND BING-JIE NI

10.1 INTRODUCTION

Large amounts of organic matter from wastewater are converted into waste activated sludge (WAS) during biological treatment processes in wastewater treatment plants (WWTPs) [1]. Anaerobic digestion of WAS has been widely applied to stabilize and reduce the volume of WAS as well as produce a renewable bioenergy resource in the form of methane [2–5]. The anaerobic digestion process generally consists four stages, i.e. hydrolysis, fermentation, acetogenesis and methanogenesis for methane production [6]. However, the application of anaerobic digestion is often limited by the slow hydrolysis rate and/or poor biochemical methane potential (or degradation extent) of the WAS [7–11].

In this work, the impacts of three different types of ZVI (i.e., iron powder, clean scrap and rusty scrap) on methane production from WAS in anaerobic digestion were evaluated systematically using both experimental and mathematical approaches. A model-based analysis was performed to reveal the mechanism of ZVI-driven enhancement of methane production from WAS. Based on the results of economic analysis, a cost-effective integrated ZVI-based anaerobic WAS digestion process was also proposed.

10.2 RESULTS

10.2.1 EFFECTS OF ZVI ADDITION ON METHANE PRODUCTION

Three types of ZVI were evaluated, i.e., iron powder, clean scrap and rusty scrap. Fig. 1 presents the methane production results from the biochemical methane potential (BMP) tests in Experiment I and II (see Methods Section). In general, ZVI addition enhanced the methane production from WAS. The increased ZVI powder addition resulted in increasing methane production (Fig. 1a). For example, 4 g/L ZVI addition increased methane production by 21% as compared to the control (0 g/L ZVI powder addition) on Day 20. As shown in Fig. 1b, 10 g/L ZVI powder, 10 g/L clean iron scrap and 10 g/L rusty iron scrap led to 11%, 22% and 30% increase on methane production, respectively, compared to that of control on Day 20. The level of methane production variation between Experimental I and II is not unexpected as real sludge from a full-scale WWTP was used, with characteristics likely varying with time. Therefore, direct comparisons between Experiment I and II are not meaningful due to the possible variation of the sludge characteristics during sludge sampling. In addition, it should be noted that the methane production by utilization of ZVI as electron donors is negligible compared to the overall methane production in the system as ZVI powder or clean scrap addition produces similar content of methane with the addition of rusty scrap.

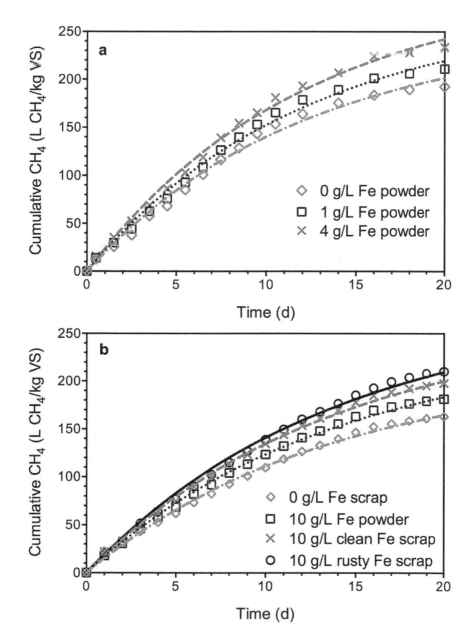

Figure 1: The measured and simulated methane production in the BMP tests (symbols represent experimental measurements and lines represent model simulations): (a) data from Experiment I; and (b) data from Experiment II.

These results demonstrated that ZVI addition indeed enhanced methane production from WAS during anaerobic digestion. The results also showed that both the clean and the rusty iron scrap were more effective than the iron powder for improving methane production from WAS. The better performance of ZVI scrap was likely due to its better contact with sludge and liquid [33]. In particular, the addition of rusty iron scrap is the most effective ZVI form for improving methane production from WAS, likely due to the fact that Fe (III) oxides on the rusty iron scrap surface could induce dissimilatory ferric iron reduction to enhance degradation of complex substrates such as WAS [34].

10.2.2 EFFECTS OF ZVI ON HYDROLYSIS RATE AND BIOCHEMICAL METHANE POTENTIAL

The hydrolysis rate (k) and biochemical methane potential (B_0) were estimated using both one-substrate and two-substrate models. Table 1 shows the estimated k and B_0 for the methane production from the WAS digestion subject to different ZVI forms and dosages using one-substrate model, while the estimated values of $k_{,rapid}$, $k_{,slow}$, $B_{0,rapid}$ and $B_{0,slow}$ in both Experiment I and II using two-substrate model are presented in Table S1 in Supplementary Material. Overall, $k_{,rapid}$ and $B_{0,rapid}$ in two-substrate model are the same as k and B_0 in one-substrate model. Both $k_{,slow}$ and $B_{0,slow}$ in two-substrate model are zero after fitting. These modeling results indicated that the WAS composition was homogeneous and the methane production from the WAS could be well described by one-substrate model.

The simulated methane production curves using one-substrate model are shown in Fig. 1, which matched all the experimental data from both Experiment I and II, further confirming the one-substrate model could well describe the methane production data. As can be seen from Table 1, the ZVI addition at all the levels applied achieved significantly higher B_0 than that of the control. The biochemical methane potential was enhanced by 9%–21% in Experiment I and 12%–29% in Experiment II compared to the corresponding control. In contrast, the ZVI addition has no effect on the k value and the obtained k values were constant in both Experiment I (ca. 0.083 d^{-1}) and Experiment II (ca. 0.072 d^{-1}) regardless of the amount of ZVI addition.

Fig. 2 shows the 95% confidence regions of k and B_0, which provide valuable information about model uncertainty and the identifiability of the obtained parameter values. The increased ZVI addition consistently resulted in better biochemical methane potential (B_0), and the confidence region moved rightward to the higher B_0 direction (x-axis) in Fig. 2. In contrast, the increased ZVI addition had no impact on the hydrolysis rate, with no real changes in confidence region locations on y-axis. In addition, there was no obvious increase in confidence region area in both Fig. 2a and 2b.

10.3 DISCUSSION

10.3.1 ZVI ADDITION IMPROVED BIOCHEMICAL METHANE POTENTIAL OF WAS RATHER THAN ITS HYDROLYSIS RATE

There are two key measures of sludge degradability that are relevant, the apparent first order degradation rate coefficient (k) and the biochemical methane potential (B_0), which represent the speed and extent of sludge conversion, respectively [35]. Model-based analysis of these two parameters and the related parameter identifiability in this work clearly showed that ZVI addition significantly enhanced methane production from WAS through improving the biochemical methane potential of WAS rather than its hydrolysis rate.

Feng et al. [32] did not look into the mechanisms for the enhanced methane production by ZVI addition and only hypothesized that the main reason might be the improved major enzyme activities related to hydrolysis and acidification. Contradictorily, this study demonstrated that the ZVI addition did not accelerate the hydrolysis rate (k) in both experiments with different types of ZVI addition. On the contrary, biochemical methane potential (B_0) was significantly improved by ZVI addition, indicating

Table 1: The estimated k and B_0 as well as the calculated Y from Experiment I and II using one-substrate model (with 95% confidence intervals)

	$k (d^{-1})$	$B_0 (L\ CH_4/kg\ VS)$	Y
Experiment I			
0 g/L Fe powder	0.083±0.007	248±12	0.44±0.02
1 g/L Fe powder	0.083±0.006	271±12	0.48±0.02
4 g/L Fe powder	0.083±0.005	300±11	0.53±0.02
Experiment II			
0 g/L Fe scrap	0.073±0.003	214±6	0.34±0.01
10 g/L Fe powder	0.072±0.003	240±5	0.37±0.01
10 g/L clean Fe scrap	0.072±0.003	262±6	0.41±0.01
10 g/L rusty Fe scrap	0.071±0.003	275±6	0.44±0.01

that ZVI increased the extent of sludge conversion and altered the sludge property [35]. It has been reported that VS destruction during anaerobic digestion of waste activated sludge generally increased with the increase of ferrous iron content in the sludge [36,37]. Indeed, ZVI can release from Fe^0 to Fe^{2+} ($Fe^0 + H^+ = Fe^{2+} + H_2$), and thus leading to a significant increase of iron content in the sludge [33,38]. As shown in Fig. 3, in this work, the released ferrous iron concentrations from ZVI also showed a good correlation with both VS reduction and the biochemical methane potential (B_0). Therefore, the alternation of sludge property to improve biochemical methane potential by ZVI could likely be the main reason for the enhanced performance of methane production.

10.3.2 A STRATEGY TO IMPLEMENT ZVI-BASED ANAEROBIC DIGESTION PROCESS IN WASTEWATER TREATMENT PLANT

From an integrated environmental and economic perspective, nutrients source in wastewater treatment systems should be managed such that both good nutrients removal performance and high resource recovery or reuse can be achieved. Based on the findings of this work, a new strategy could be proposed to simultaneously enhance methane production from WAS and iron resource reuse through integrating the ZVI-based anaerobic digestion process of this work with the conventional chemical phosphorus removal process in WWTPs.

As presented in Fig. 4, waste iron scrap (the most efficient ZVI as demonstrated in this work) can be freely obtained from machinery factory and then transported to the WWTP. The obtained iron scrap (ZVI) can be added to the anaerobic digester in order to enhance the methane production by increasing the biochemical methane potential. In anaerobic digester, ZVI can be released from Fe^0 to Fe^{2+}, and thus eliminated the potential sulfide production/accumulation issues as well as the possible H_2S emission in the biogas in traditional anaerobic digester through iron sulfide precipitation [39]. This in turn could further enhance the performance of WAS digestion without additional chemical cost from external ferrous/ferric iron dosing [40]. With regard to the generation of organic sulfur odors from the dewa-

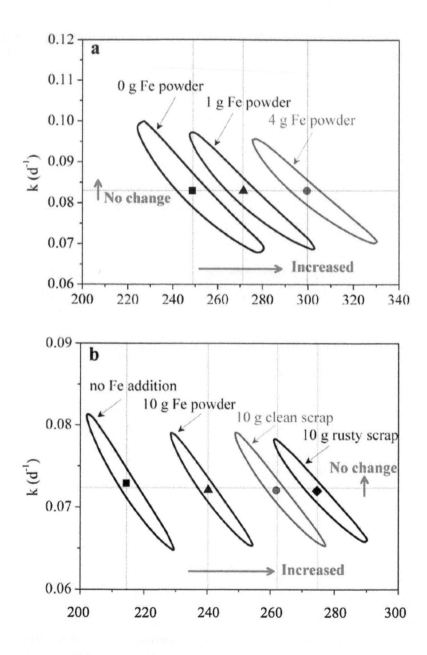

Figure 2: The 95% confidence regions of the estimated hydrolysis rate (k) and biochemical methane potential (B_0) with different ZVI additions: (a) using data from Experimental I; and (b) using data from Experiment II.

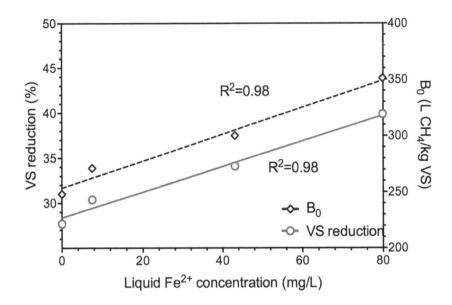

Figure 3: Relationships between the released ferrous iron concentrations and the percentage of VS reduction as well as the obtained B_0 value in Experiment I.

tered sludge cakes, iron could also reduce odor-causing gases, resulting in better quality of dewatering sludge. More importantly, the Fe (II) in anaerobic digestion liquor can be recycled to bioreactor and further oxidized to Fe (III), which can be used for chemical phosphors removal via the generation of $FePO_4^{4+}$. This strategy would not only represent a significant process cost reduction (further discussed below), but also improve the sludge and wastewater treatment efficiency, enabling maximum resource (iron) reuse while achieving improved methane production. In addition, from a network-wide view, commonly used ferric iron dosing in sewers for H_2S control [42] might also be useful for CH_4 production enhancement during anaerobic digestion and phosphors removal in the WWTP.

10.3.3 POTENTIAL ECONOMIC FEASIBILITY OF ZVI-BASED TECHNOLOGY FOR ENHANCING BIOLOGICAL METHANE PRODUCTION

It has been demonstrated that the estimated lab-scale BMP results are more conservative or comparable to full-scale test results [43]. Thus, the

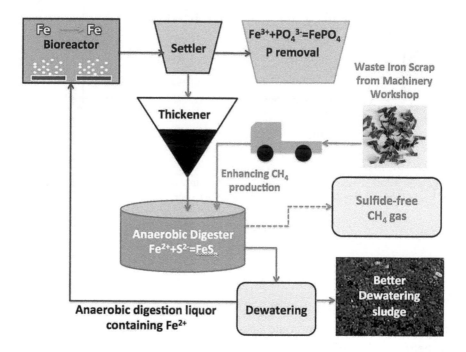

Figure 4: A proposed strategy to integrate ZVI-based anaerobic digestion process of this work with the conventional chemical phosphorus removal process in wastewater treatment plant. ZVI addition in anaerobic digester can enhance methane production from WAS. The sulfide produced in anaerobic digester can be precipitated by ferrous iron that produced from ZVI addition, resulting in enhanced sulfide-free biogas (methane) production. The anaerobic digestion liquor containing Fe (II) can be reused and fed into bioreactors, in which the Fe (II) can be oxidized to Fe (III). The generated Fe (III)-containing effluent can then be used for chemical phosphorus removal process, to form a cost-effective and environment-friendly technology, enabling maximum resource recovery/ reuse while achieving enhanced methane production in wastewater treatment system.

estimated values obtained in the current study are used for a conservative assessment of the potential economic feasibility of the proposed ZVI-based anaerobic digestion technology. This was carried out by a desktop scaling-up study on a full-scale WWTP with a population equivalent (PE) of 400,000 and with an anaerobic sludge digester at a hydraulic retention time (HRT) of 20 days. 10 g/L rusty iron scrap was chosen for the following economic evaluations.

From the Fe^{2+} released (41 mg/L), theoretically, the iron scrap could be recycled for approximately 243 batches (10*1000/41) if the loss of iron solid through effluent is ignored. With a 29% increase in methane production at this level of ZVI addition, the net economic benefit is estimated to be around $231,000 per annum compared with the system without ZVI addition (see Table S2 in Supplementary Material). The net benefit arises from the enhanced methane production associated benefit (i.e., its conversion to heat and power) ($150,000 per annum) and decreased WAS transport and disposal costs ($90,000 per annum) overweighing the additional costs for ZVI transport and ZVI chamber ($9,000 per annum). The advantages of ZVI addition on sulfide control in digester, phosphors removal through anaerobic digestion liquor recycle and better dewatering sludge have not been considered. Therefore, the ZVI-based technology is potentially economically attractive indeed. However, the benefit and cost values presented should be considered indicative only. In particular, they may vary from region to region and from country to country, depending on the local conditions. In addition, the direct quantitative economic and performance comparison with other available technologies are difficult at this stage since the results depend on many factors including the WAS characteristics among others [23], which remains further investigations in the future. This could and should be done in future studies by performing experiments using the same WAS and under similar operating conditions.

Moreover, it should be noted that there is no environmental consequence of the proposed chemical-free ZVI-based technology based on CO_2 emission, revealing this approach being environmental friendly. In comparison, some other WAS pretreatment technologies (i.e., thermal and alkaline pretreatment) might cause negative environmental effect [44]. Different from temperature phased anaerobic digestion and mechanical pretreatment which generally increase k [45,46], this ZVI-based approach improved B_0, thus potentially allowing more methane production in terms of performance improvements for anaerobic digesters. Since it does not improve the degradation rate, this requires the same amount of HRT in order to achieve maximized sludge reduction. It should be noted that lab-scale batch tests were performed in our study. Full-scale system may behave differently in terms of k and B_0 as demonstrated in Bastone et al. [43]. Therefore, full-scale trials are needed to further evaluate this technology.

In summary, the effects of three different types of ZVI (i.e., iron powder, clean scrap and rusty scrap) on methane production from WAS in anaerobic digestion were investigated by using both experimental and mathematical approaches. The results demonstrated that both the clean and the rusty iron scrap were more effective than the iron powder for improving methane production from WAS. ZVI addition significantly enhanced methane production from WAS through improving the biochemical methane potential of WAS rather than its hydrolysis rate. The alternation of sludge property by ZVI resulted in improved biochemical methane potential and thus the enhanced methane production. The ZVI-based anaerobic digestion process could be potentially implemented and integrated with the conventional chemical phosphorus removal process in wastewater treatment plant to form a cost-effective and environment-friendly technology, enabling maximum resource recovery/reuse while achieving enhanced methane production in wastewater treatment system.

10.4 METHODS

10.4.1 WASTE ACTIVATED SLUDGE

The waste activated sludge used in this work was collected from the sludge treatment unit at a full-scale municipal wastewater treatment plant in Dalian, China. The sludge was stored at 4 °C before use. The volatile solids

(VS) to total chemical oxygen demand (TCOD) ratios of the sludge used for methane production ranged between 0.60 and 0.67.

10.4.2 ZVI SOURCES

Three types of ZVI were evaluated, i.e., iron powder, clean scrap and rusty scrap. The ZVI powder has a diameter of 0.2 mm with BET surface area of 0.05 m^2/g and purity >98%. The rusty scrap (about 8 mm * 4 mm * 0.5 mm, purity >95%) was obtained from a machinery workshop in Dalian, China. The clean scrap was acquired through a pretreatment of the rusty scrap to remove the rusty cover. The difference between the two scraps is that the rusty scrap had a corrosion layer covering the surface of the scrap [33].

10.4.3 ANAEROBIC BIOCHEMICAL METHANE POTENTIAL TESTS

In order to evaluate the effect of different forms of ZVI on methane production in anaerobic digestion, methane production from the WAS with different types of ZVI addition was assessed using anaerobic batch BMP tests [32]. The inoculum for the BMP tests was collected from an anaerobic digester [33]. Two types of batch experiments were performed. In Experiment I, 0, 1.0, and 4.0 g/L of ZVI powder were added into three identical sets of BMP test vials, respectively. In Experiment II, 10 g/L ZVI powder, 10 g/L clean scrap and 10 g/ L rusty scrap were used as ZVI sources and dosed to three identical sets of BMP vials for comparison, with a control test in which no ZVI was added.

In each test, WAS, ZVI and the inoculum obtained from the anaerobic digester were added into serum vials for BMP tests. After that, the vials were capped with silica gel stoppers. The oxygen was removed from the headspace by exchanging it with nitrogen gas for at least 10 min. All BMP tests were conducted at 35±1 °C for 20 d. The biogas (methane) production in BMP vials was collected and monitored by using gas chromatograph (Shimadzu, GC-14C) equipped with a thermal conductivity detector. More details of the BMP tests can be found elsewhere [32,33].

10.4.4 MODEL-BASED ANALYSIS

The hydrolysis rate (k) and biochemical methane potential (B_0) are the two key parameters associated with methane production from WAS [7,8,10]. In this work, these two parameters were used to evaluate and compare the methane production kinetics and potential of the WAS at different ZVI levels or with different types of ZVI. They were estimated by fitting the methane production data from the BMP tests to a first-order kinetic model using a modified version of Aquasim 2.1d with sum of squared errors (J_{opt}) as an objective function [43]. The uncertainty surfaces of k and B_0, based on a model-validity statistical F-test with 95% confidence limits, were also estimated by using Aquasim 2.1d [43].

Two models were applied. The first one considered a single substrate type (i.e., one- substrate model) in the first-order kinetic model [22,43], as shown in Equation (1):

$$B(t)=B_0(1-e^{-kt})$$

where B(t) (L CH_4/kg VS) is the cumulative methane production at time t(d).

In the second model, the WAS samples comprised a rapidly biodegradable substrate type and a slowly biodegradable substrate type (i.e. two-substrate model) [47]. The equation of the two-substrate model is shown below:

$$B(t)=B_{0,rapid}(1-e^{-k_{rapid}t})+B_{0,slow}(1-e^{-k_{slow}t})$$

where $B_{0,rapid}$ and $B_{0,slow}$ (L CH_4/kg VS) are biochemical methane potentials of the rapidly biodegradable substrates and slowly biodegradable substrates, respectively; k_{rapid} and k_{slow} (d^{-1}) are hydrolysis rates of the rapidly biodegradable substrates and slowly biodegradable substrates, respectively.

Based on the determined B_0, the degradation extent (Y) of WAS could then be calculated using Equation (3):

$$Y=B_0/380 \times R_{WAS}$$

where 380 (L CH_4/kg TCOD) is theoretical biochemical methane potential of WAS under standard conditions (25 °C, 1 atm); RWAS is the measured VS to TCOD ratio in the WAS.

REFERENCES

1. Duan, N., Dong, B., Wu, B. & Dai, X. High-solid anaerobic digestion of sewage sludge under mesophilic conditions: Feasibility study. Bioresour. Technol. 104, 150–156 (2012).
2. Eskicioglu, C., Prorot, A., Marin, J., Droste, R. L. & Kennedy, K. J. Synergetic pretreatment of sewage sludge by microwave irradiation in presence of H_2O_2 for enhanced anaerobic digestion. Water Res. 42, 4674–4682 (2008).
3. Tiehm, A., Nickel, K., Zellhorn, M. & Neis, U. Ultrasonic waste activated sludge dis-integration for improving anaerobic stabilization. Water Res. 35, 2003–2009 (2001).
4. Wang, L., Aziz, T. N. & de los Reyes, F. L. Determining the limits of anaerobic co-digestion of thickened waste activated sludge with grease interceptor waste. Water Res. 47, 3835–3844 (2013).
5. McCarty, P. L., Bae, J. & Kim, J. Domestic wastewater treatment as a net energy producer–can this be achieved? Environ. Sci. Technol. 45, 7100–7106 (2011).
6. Lv, W., Schanbacher, F. L. & Yu, Z. Putting microbes to work in sequence: Recent advances in temperature-phased anaerobic digestion processes. Bioresour. Technol. 101, 9409–9414 (2010).
7. Wang, Q. et al. Free nitrous acid (FNA)-based pretreatment enhances methane production from waste activated sludge. Environ. Sci. Technol. 47, 11897–11904 (2013).
8. Labatut, R. A., Angenent, L. T. & Scott, N. R. Biochemical methane potential and biodegradability of complex organic substrates. Bioresour. Technol. 102, 2255–2264 (2011).
9. Hosseini Koupaie, E., Barrantes Leiva, M., Eskicioglu, C. & Dutil, C. Mesophilic batch anaerobic co-digestion of fruit-juice industrial waste and municipal waste sludge: Process and cost-benefit analysis. Bioresour. Technol. 152, 66–73 (2014).
10. Lissens, G., Thomsen, A. B., De Baere, L., Verstraete, W. & Ahring, B. K. Thermal wet oxidation improves anaerobic biodegradability of raw and digested biowaste. Environ. Sci. Technol. 38, 3418–3424 (2004).
11. Mu, Y., Yu, H.-Q. & Wang, G. A kinetic approach to anaerobic hydrogen- producing process. Water Res. 41, 1152–1160 (2007).

12. Chu, L., Yan, S., Xing, X.-H., Sun, X. & Jurcik, B. Progress and perspectives of sludge ozonation as a powerful pretreatment method for minimization of excess sludge production. Water Res. 43, 1811–1822 (2009).
13. Imbierowicz, M. & Chacuk, A. Kinetic model of excess activated sludge thermohydrolysis. Water Res. 46, 5747–5755 (2012).
14. Ibeid, S., Elektorowicz, M. & Oleszkiewicz, J. A. Modification of activated sludge properties caused by application of continuous and intermittent current. Water Res. 47, 903–910 (2013).
15. Vlyssides, A. G. & Karlis, P. K. Thermal-alkaline solubilization of waste activated sludge as a pre-treatment stage for anaerobic digestion. Bioresour. Technol. 91, 201–206 (2004).
16. Nah, I. W., Kang, Y. W., Hwang, K.-Y. & Song, W.-K. Mechanical pretreatment of waste activated sludge for anaerobic digestion process. Water Res. 34, 2362–2368 (2000).
17. Zhang, D., Chen, Y., Zhao, Y. & Zhu, X. New sludge pretreatment method to improve methane production in waste activated sludge digestion. Environ. Sci. Technol. 44, 4802–4808 (2010).
18. Zhang, D., Chen, Y., Zhao, Y. & Ye, Z. A new process for efficiently producing methane from waste activated sludge: alkaline pretreatment of sludge followed by treatment of fermentation liquid in an EGSB reactor. Environ. Sci. Technol. 45, 803–808 (2010).
19. Mehdizadeh, S. N., Eskicioglu, C., Bobowski, J. & Johnson, T. Conductive heating and microwave hydrolysis under identical heating profiles for advanced anaerobic digestion of municipal sludge. Water Res. 47, 5040–5051 (2013).
20. Eskicioglu, C., Terzian, N., Kennedy, K. J., Droste, R. L. & Hamoda, M. Athermal microwave effects for enhancing digestibility of waste activated sludge. Water Res. 41, 2457–2466 (2007).
21. Mu, Y., Zheng, X.-J., Yu, H.-Q. & Zhu, R.-F. Biological hydrogen production by anaerobic sludge at various temperatures. Int. J. Hydrogen Energy 31, 780–785 (2006).
22. Wang, Q., Jiang, G., Ye, L. & Yuan, Z. Enhancing methane production from waste activated sludge using combined free nitrous acid and heat pre-treatment. Water Res. 63, 71–80 (2014).
23. Carre`re, H. et al. Pretreatment methods to improve sludge anaerobic degradability: a review. J. Hazard. Mater. 183, 1–15 (2010).
24. Fu, F., Dionysiou, D. D. & Liu, H. The use of zero-valent iron for groundwater remediation and wastewater treatment: A review. J. Hazard. Mater. 267, 194–205 (2014).
25. Mu, Y., Yu, H.-Q., Zheng, J.-C., Zhang, S.-J. & Sheng, G.-P. Reductive degradation of nitrobenzene in aqueous solution by zero-valent iron. Chemosphere 54, 789–794 (2004).
26. Zhang, J. et al. Bioaugmentation and functional partitioning in a zero valent iron-anaerobic reactor for sulfate-containing wastewater treatment. Chem. Eng. J. 174, 159–165 (2011).
27. Liu, Y. et al. Effects of an electric field and zero valent iron on anaerobic treatment of azo dye wastewater and microbial community structures. Bioresour. Technol. 102, 2578–2584 (2011).

28. Zhang, Y., Liu, Y., Jing, Y., Zhao, Z. & Quan, X. Steady performance of a zero valent iron packed anaerobic reactor for azo dye wastewater treatment under variable influent quality. Journal of Environmental Sciences 24, 720–727 (2012).

29. Zhang, Y., Jing, Y., Quan, X., Liu, Y. & Onu, P. A built-in zero valent iron anaerobic reactor to enhance treatment of azo dye wastewater. Water Sci. Technol. 63, 741–746 (2011).

30. Liu, Y. et al. Optimization of anaerobic acidogenesis by adding Fe0 powder to enhance anaerobic wastewater treatment. Chem. Eng. J. 192, 179–185 (2012).

31. Liu, Y. et al. Enhanced azo dye wastewater treatment in a two-stage anaerobic system with Fe0 dosing. Bioresour. Technol. 121, 148–153 (2012).

32. Feng, Y., Zhang, Y., Quan, X. & Chen, S. Enhanced anaerobic digestion of waste activated sludge digestion by the addition of zero valent iron. Water Res. 52, 242–250 (2014).

33. Zhang, Y., Feng, Y., Yu, Q., Xu, Z. & Quan, X. Enhanced high-solids anaerobic digestion of waste activated sludge by the addition of scrap iron. Bioresour. Technol. 159, 297–304 (2014).

34. Lovley, D. R. Organic matter mineralization with the reduction of ferric iron: a review. Geomicrobiol. J. 5, 375–399 (1987).

35. Ge, H., Jensen, P. D. & Batstone, D. J. Increased temperature in the thermophilic stage in temperature phased anaerobic digestion (TPAD) improves degradability of waste activated sludge. J. Hazard. Mater. 187, 355–361 (2011).

36. Novak, J., Verma, N. & Muller, C. The role of iron and aluminium in digestion and odor formation. Water Sci. Technol. 56, 59–65 (2007).

37. Park, C., Muller, C. D., Abu-Orf, M. M. & Novak, J. T. The effect of wastewater cations on activated sludge characteristics: effects of aluminum and iron in floc. Water Environ. Res 78, 31–40 (2006).

38. Liu, Y., Zhang, Y., Quan, X., Chen, S. & Zhao, H. Applying an electric field in a built-in zero valent iron – Anaerobic reactor for enhancement of sludge granulation. Water Res. 45, 1258–1266 (2011).

39. Chen, Y., Cheng, J. J. & Creamer, K. S. Inhibition of anaerobic digestion process: A review. Bioresour. Technol. 99, 4044–4064 (2008).

40. Park, C. M. & Novak, J. T. The effect of direct addition of iron (III) on anaerobic digestion efficiency and odor causing compounds. Water Sci. Technol. 68, 2391–2396 (2013).

41. Gutierrez, O., Park, D., Sharma, K. R. & Yuan, Z. Iron salts dosage for sulfide control in sewers induces chemical phosphorus removal during wastewater treatment. Water Res. 44, 3467–3475 (2010).

42. Ganigue, R., Gutierrez, O., Rootsey, R. & Yuan, Z. Chemical dosing for sulfide control in Australia: an industry survey. Water Res. 45, 6564–6574 (2011).

43. Batstone, D. J., Tait, S. & Starrenburg, D. Estimation of hydrolysis parameters in full-scale anerobic digesters. Biotechnol. Bioeng. 102, 1513–1520 (2009).

44. Carballa, M., Duran, C. & Hospido, A. Should we pretreat solid waste prior to anaerobic digestion? An assessment of its environmental cost. Environ. Sci. Technol. 45, 10306–10314 (2011).

45. Ge, H., Jensen, P. D. & Batstone, D. J. Temperature phased anaerobic digestion increases apparent hydrolysis rate for waste activated sludge. Water Res. 45, 1597–1606 (2011).
46. Donoso-Bravo, A., Pe´rez-Elvira, S. I. & Fdz-Polanco, F. Application of simplified models for anaerobic biodegradability tests. Evaluation of pre-treatment processes. Chem. Eng. J. 160, 607–614 (2010).
47. Rao, M. S., Singh, S. P., Singh, A. K. & Sodha, M. S. Bioenergy conversion studies of the organic fraction of MSW: assessment of ultimate bioenergy production potential of municipal garbage. Applied Energy 66, 75–87 (2000).

There are several supplemental files that are not available in this version of the article. To view this additional information, please use the citation on the first page of this chapter.

CHAPTER 11

Towards a Metagenomic Understanding on Enhanced Biomethane Production from Waste Activated Sludge after pH 10 Pretreatment

MABEL TING WONG, DONG ZHANG, JUN LI,
RAYMOND KIN HI HUI, HEIN MIN TUN, MANREETPAL SINGH
BRAR, TAE-JIN PARK, YINGUANG CHEN AND FREDERICK C LEUNG

11.1 BACKGROUND

Activated sludge technology is currently the most broadly-implemented biological method for biomass conversion in wastewater treatment plants (WWTPs) [1]. However, vast quantities of highly organic waste activated sludge (WAS) is produced during this process, and this by-product mass continues to increase with the expansion of population and industry [1-4]. Sludge disposal by landfill or incineration may no longer be feasible in the near future due to land scarcity, high waste charge and increasingly stringent environmental control regulations [4,5]. As a result, the strategy for sludge management is shifting towards its re-utilization as a potential source for renewable energy [6-8]. In this regard, the anaerobic digestion process represents an attractive means of sludge reduction while producing renewable energy in the form of biogas [8,9]. Identifying efficient ways to improve methane production, a major biogas product from anaerobic digestion, has now become a topic of interest for numerous researchers [3,10,11].

The performance of an anaerobic digestion system has been shown to be tied closely to its microbial community structure [12]. Methane production from WAS is a complex, multi-step process which involves multiple syntrophic interactions within the microbial consortium [13]. Complex compounds (polysaccharides, proteins, nucleic acids, and lipids) are first converted to oligomers and monomers through the action of extracellular hydrolytic enzymes produced by the primary fermenting bacteria [14]. Subsequently, the intermediate products are further transformed into acetate, carbon dioxide, hydrogen and formate by secondary fermenters [14]. The final methanogenesis step is then conducted by methanogenic archaea, whose energy substrates are highly restricted to acetate, H_2, CO_2, formate or certain C_1 compounds [15]. To improve methane yield from sludge, enormous research efforts have been devoted to the development of pretreatment methods to accelerate sludge hydrolysis, including thermal [16], thermal-alkaline [17], ultrasonic [18], mechanical and thermo-chemical methods [19]. Although the field of pretreatment research has progressed significantly in the past decade, many significant questions related to their effects on the underlying microbial interactions remain unanswered.

As the field of pretreatment method research is nearing a threshold, the accomplishments of the past are pushing on the door of microbiology to provide new insights [20,21]. In this study, we aimed to investigate the microbial composition and gene content of sludge subjected to our novel pretreatment method (maintaining pH at 10 for 8 days) which leads to significantly enhanced methane generation compared to other documented methods (ultrasonic, thermal and thermal-alkaline) [4]. It was reported in our previous study that both sludge hydrolysis and short-chain fatty acids (SCFAs; eg. acetic, butyric and propionic acid) accumulation were significantly enhanced when WAS was anaerobically fermented under the condition of pH 10 for 8 days [22]. This phenomenon was suspected to be due to biotic factors rather than abiotic ones (e.g. alkaline hydrolysis) since much higher SCFAs accumulation and enzyme activities were observed in un-autoclaved sludge compared to autoclaved sludge [22]. Further investigation by our group suggested that the solubilization of the sludge matrix, usually a hydrolysis event by the embedded extracellular enzymes, may contribute to the significant SCFAs improvement [23].

In this study, a shotgun metagenomic approach was chosen to study potential shifts in microbial communities and/or gene contents that could help explain elevated productions of methane under our novel pretreatment method [24,25]. The latest advances in pyrosequencing technology afford new opportunities to undertake such metagenomic studies to explore the dynamics of microbial communities in time, space or under fluctuating environmental conditions with un-precedented levels of microbial diversity coverage and depth [26-28]. In addition to elucidating the microbiology underpinning the sludge pretreatment process [29], our study sought to improve knowledge of the diversity and physiology of participating syntrophs and methanogens, as well as the mechanism behind the establishment and maintenance of mutualistic cooperation. This knowledge will help to establish a better control over the hydrolysis and methanogenic processes, and promote pretreatment as part of a pertinent strategy for sludge management in WWTP. To the best of our knowledge, this study is the first to adopt a whole genome shotgun approach to link the knowledge between microbiology and engineering of sludge pretreatment methods. Additional statistical soundness was conferred to the surmised conclusion, as the comparative analysis was conducted between meta-datasets generated by the same sequencing method, from well-characterized experimental designs which only differ in known parameters by manipulation.

11.2 RESULTS AND DISCUSSION

11.2.1 ENHANCED METHANE PRODUCTION BY USING A NEWLY DEVISED PRETREATMENT STRATEGY

Collected WAS was subjected to a novel pretreatment method (maintaining pH 10 for 8 days) [4], and its effects on methane production enhancement was characterized in this study. Figure 1 illustrates the experimental flow- chart of obtaining end samples from the biogas-producing anaerobic digesters. Pretreated sludge for 30 days (P30), un-pretreated sludge for 30 days (UP30), or 40 days (UP40) were used as substrate. Pretreated sludge was significantly more effective in producing biogas, as total gas volume produced on day 8 (604.32 ml/g VSS-added) is 7.71 and 1.67 times of that in the un-pretreated sludge bioreactor on day 8 (78.43 ml/g VSS-added) and day 16 (360.99 ml/g VSS-added), respectively (Figure 2A-B). Methane production was 389.8 mL-CH_4/g VSS-added using pretreated sludge as a substrate, which was 3.78 times that of un-pretreated sludge (103.2 mL-CH_4/g VSS-added) (see Figure 2C). The generated methane on average constituted 70.5% and 59.1% of the total generated gas composition in the two respective bioreactors, while the rest was primarily CO_2 (pretreated sludge, 3.6%; un-pretreated sludge, 9.2%) and other small amounts of N_2, H_2, NH_3 and H_2S (data not shown) (Figure 2D-E). It is worthwhile to note that the methane content was increased by 19.3%, and carbon dioxide content was decreased by 60.9% after the pretreatment. The reproducibility of this new pretreatment strategy validated the experimental success seen before and its technical robustness. This warranted investigation of biological mechanisms behind this significant bioreactor improvement in performance.

11.2.2 BIOGAS PRODUCING MICROBIAL COMMUNITY RESIDING IN ANAEROBIC DIGESTER INFERRED BY METAGENOME SEQUENCING

To analyze biogas-producing microbes residing in our studied bioreactors in terms of community structure, gene content, metabolic capabilities and the role of specific organisms in biogas formation, a metagenomic approach using the 454 pyrosequencing was used. Statistical data summarizing the output sequencing quantity of the three independent runs of P30, UP30 and UP40 is given in Table 1. To detect differentially abundant features between the microbial communities, we used the Meta Genome Rapid Annotation using Subsystem Technology (MG-RAST) analysis pipeline, which has shown to be applicable to various metagenomic data performing taxonomic classification as well as functional annotation [30]. Composition of the biogas-producing microbial community was obtained by taxonomic classification of reads base on the M5NR database on the

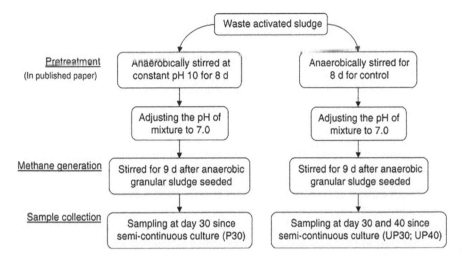

Figure 1: Sample preparation (P30, UP30 and UP 40) from pretreated and un-pretreated sludge bioreactors.

MG-RAST platform. In all three reactor samples, Bacteria were the dominant superkingdom (Table 1). Domain-based allocation was primarily assigned to Bacteria (82.13% on average) and Archaea (11.86% on average), whilst the assigned reads to Eukaryota and Viruses altogether accounted for less than half a percent (Table 1). The bacterial proportion of the microbial community in P30 was the highest (85.46%) followed by UP30 (81.82%) and UP40 (79.10%). As for the Archaea domain, it ranged from 8.32% in P30, to 12.13% in UP30, and to 15.13% in UP40. The clustering pattern shown in Additional file 1 provides a graphical representation of the overall taxonomic similarity—the microbiomes of the unpretreated sludge bioreactors (UP30 and UP40) displayed high resemblance, in which the mature unpretreated sludge bioreactor (UP40, operated for longer time) is comparatively closer to the significantly diverse pretreated sludge bioreactor. The result provided the first insight into both the temporal dynamic and the enormous impact of pretreatment on microbiology. Downstream analyses focused on the Bacteria and Archaea domains (unless otherwise specified, the percentages below are representative of the identified reads within each domain per independent run).

Within the Bacteria domain, the top abundant phyla were Proteobacteria (UP30, 32.17%; UP40, 30.45%; P30, 42.28%), Bacteroidetes (UP30, 22.91%; UP40, 22.25%; P30, 23.38%), Firmicutes (UP30, 16.54%; UP40, 17.65%; P30, 10.97%), Actinobacteria (UP30, 7.13%; UP40, 6.94%; P30, 6.83%) and Chloroflexi (UP30, 5.00%; UP40, 5.21%; P30, 3.37%), they collectively account for over 0.7 of the bacterial reads (normalized between 0 and 1) for each of the three bioreactors (Figure 3A). Proteobacteria, Bacteroidetes and Firmicutes have been reported to be dominant phyla as well in similar analysis of anaerobic digestion of sludge [9]. Further resolution at the class level revealed that the microbial compositions overlapped between the bioreactor samples. To be more precise, the dominant bacterial lineages presented in this study were related to anaerobic digestion, including Bacteroidia (UP30, 13.50%; UP40, 12.69%; P30, 8.90%), δ-proteobacteria (UP30, 13.08%; UP40, 12.71%; P30, 8.93%), Clostridia (UP30, 11.42%; UP40, 12.21%; P30, 7.31%), γ-proteobacteria (UP30, 8.12%; UP40, 7.90%; P30, 11.18%), Actinobacteria (class) (UP30, 7.37%; UP40, 7.20%; P30, 7.06%), α-proteobacteria (UP30, 5.78%; UP40, 5.28%; P30, 9.33%), β-proteobacteria (UP30, 5.73%; UP40, 4.86%; P30, 13.58%), Bacilli (UP30, 5.16%; UP40, 5.25%; P30, 3.53%), Flavobacteria (UP30, 4.70%; UP40, 4.76%; P30, 6.98%) and Chloroflexi (class) (UP30, 2.65%; UP40, 2.56%; P30, 1.84%) (Figure 3B).

With time, there was moderate temporal variation in the relative abundances of these bacterial members in the anaerobic digester fed with un-pretreated sludge (UP30 and UP40). While the relative proportion of these communities remained comparable between UP30 and UP40, a larger extent of differences could be observed when P30 was taken into comparison for certain lineages. The over-represented bacterial members in P30 included α-proteobacteria (9.33%), β-proteobacteria (13.58%) and γ- proteobacteria (11.18%); which are respectively 69.06%, 158.21% and 39.60% more than UP30 and UP40 on average. Overall, these sequences were originated from the WAS substrate, AGS inoculum and the biogas-producing symbiont within the system, hence the differences observed in their relative abundance were attributable to the sole variable of either using pretreated or un-pretreated sludge in the bioreactors. These results suggested that the bacterial communities that underlie the anaerobic di-

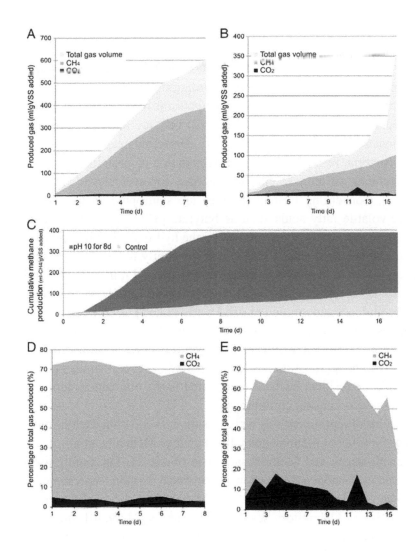

Figure 2: Biogas production profile of pretreated and un-pretreated sludge bioreactors. Efficiency of biogas production in (A) the pretreated sludge bioreactor is significantly higher than (B) the un-pretreated sludge bioreactor, giving rise to a 7.71-fold difference in the produced volume at day 8. (C) The cumulative methane production was 389.8 mL-CH_4/g VSS-added using pretreated sludge as substrate, 3.78 times of that with un-pretreated sludge (103.2 mL-CH_4/g VSS-added) as substrate. Figure 2 (D-E) display methane and carbon dioxide percentages produced by (D) pretreated and (E) un-pretreated sludge bioreactor, respectively; while methane is the major biogas component in both reactors, the methane content was increased by 19.3% with a decrease in carbon dioxide content by 60.9% after pretreatment.

gesters were dynamic, and they responded rapidly to the pretreated sludge substrate and change substantially over time.

Based on the phylogenetic affiliation of the metagenomic sequences, it is possible to form hypotheses regarding the metabolic functions of the groups [9]. Proteobacteria are microorganisms involved in the initial steps of degradation, studies have shown that they are the main consumers of propionate, butyrate and acetate [31]. Bacteroidetes are known proteolytic bacteria, responsible for the degradation of protein and subsequent fermentation of amino acids into acetate for acetoclastic methanogens [31,32]. Concerning Firmicutes, they are syntrophic bacteria which degrade volatile fatty acids such as butyrate [31]. The H_2 generated from this process could be then uptaken by the hydrogenotrophic methanogens [33]. The bacterial class Clostridia is frequently found in co-culture with other species in biomasss conversion systems and it includes a number of anaerobic species that are commonly associated with the decomposition of lignocelluloses and municipal solid waste [34,35]. As for Chloroflexi, this group is often found in various wasterwater treatments such as anaerobic digesters and biological nutrient removal processes; their potential role in carbohydrate degradation has been reported in several studies [31,36-38]. Overall, the resident bacteria manifested in the bioreactors represented a biomass decomposing community.

Archaeal representatives were less diverse and consisted of only three major families of methanogens, belonging to Methanosarcinaceae, Methanobacteriaceae and Methanosaetaceae (Table 2). The predominance of selected methanogen lineages in the Archaeal domain (represented as percentage of total identified reads in Archaea) has been observed in both production and laboratory-scale biogas reactors [23,39,40], and concluded in a recent meta-analysis of the collective microbial diversity [41,42]. At the genus level, *Methanosarcina* was determined to be the most abundant methanogen in all three bioreactors (13.51% on average—UP30, 15.10%; UP40, 13.32%; P30, 12.10%), while *Methanosaeta* (10.67% on average—UP30, 13.13%; UP40, 10.31%; P30, 8.56%) and *Methanothermobacter* (10.35% on average— UP30, 10.73%; UP40, 9.41%; P30, 10.91%) contributed to slightly lesser portions. Anaerobic digesters are typical habitats to these three genera of methanogens [43], and the enrichment of *Methanosarcina* species was in congruence with other evaluation studies on

Table 1: Statistics of 454 GS Junior pyrosequence datasets presented in this study

Sample	P30	UP30	UP40
Pretreatment	pH 10 for 8d	Un-pretreated	Un-pretreated
Semi-continuous culture	30d	30d	40d
Sequences count	151,676	114,694	134,522
Mean sequence length	392±120 bp	289±99 bp	425±115 bp
Archaea	8.32%	12.13%	15.13%
Bacteria	85.46%	81.82%	79.10%
Eukaryota	0.45%	0.30%	0.48%
Viruses	0.01%	0.02%	0.02%
other sequences	0.01%	0.01%	0.00%
unassigned	5.75%	5.72%	5.28%

Legends: By taxonomic classification based on M5NR database, Bacteria (82.13% on average) and Archaea (11.86% on average) are the dominant domains, with variable proportions in the bioreactor samples. UP30 and UP40 are collected from same reactor at different time since semi-continuous culture.

primary sludge and WAS anaerobic digesters [44,45]. The observation of a lesser portion of *Methanothermobacter* was expected, as the operating condition of the bioreactor (approximate 35°C) was not in favour of this thermophilic Archeon's proliferation (55-65°C) [43].

Methanosaeta species were observed to be more abundant in samples of bioreactor digesting unpretreated sludge (UP30; 13.13%) than that with pretreated sludge (P30; 8.56%), but the population decreases with time (UP40; 8.56%), A similar observation was found in an earlier surveillance of the methanogenic population dynamics in anaerobic digesters, where the hitherto abundant *Methanosaeta* population decreased rapidly as the acetate concentration increased [12]. As acetoclastic methanogens, both *Methanosarcina* and *Methanosaeta* are able to split acetate, oxidize the carboxyl-group to CO_2 and reduce the methyl group to CH_4 [43]. At the same time, while *Methanosarcina* thrive in environments with high acetate concentration; lower acetate concentrations benefit the dominance of *Methanosaeta* owing to their high affinity for acetate [12,46-48]. As aforementioned, an enhanced bioproduction of acetic acid was reported in the pretreated sludge [4], the revealed shift in the methanogenic community could therefore be interpreted as an ecological consequence [30,49,50] of the sludge pretreatment.

11.2.3 METHANE PRODUCTION VIA ACETOCLASTIC AND HYDROGENOTROPHIC PATHWAYS

To identify the methane-producing organisms in the bioreactor samples, taxonomic and functional affiliation of the metagenomic reads were evaluated in parallel (represented as percentage of total identified methanogens). On the species level, it is common to all three bioreactor samples that *Methanosaeta thermophila* (12.36% on average—UP30, 14.48%; UP40, 12.36%; P30, 10.25%) and *Methanothermobacter thermautotrophicus* (9.34% on average—UP30, 9.50%; UP40, 8.59%; P30, 9.94%) are the over-represented methanogens (Table 2). Here, *Methanosarcina* species were not represented as one of the top dominant methanogens (13.51% on average— UP30, 15.10%; UP40, 13.32%; P30, 12.10%), this observation was linked to the intra-divergence of each family [9]. *M. thermophila* is an obligate methanogen which consumes acetate only [51], whereas *M. thermautotrophicus* conserves energy by using H_2 to reduce CO_2 to CH_4 [33,52]. While the pH was adjusted to be within the optimum range (7.0) for the two dominate methanogens after pretreatment (Figure 1), it was interesting to detect the prevalence of the thermophilic *M. thermautotrophicus* in the bioreactors that operated at mesophilic temperature [53]. At the same time, functional enzyme-encoding genes for the two methanogenesis pathways were identified with reference to KEGG and Metacyc pathway database entries (Figure 4, Additional file 2) [54,55]. Nonetheless, based on these results, it was evinced that the production of methane in the studied bioreactors was performed by methanogenic Archaea via acetoclastic and hydrogenotrophic pathways.

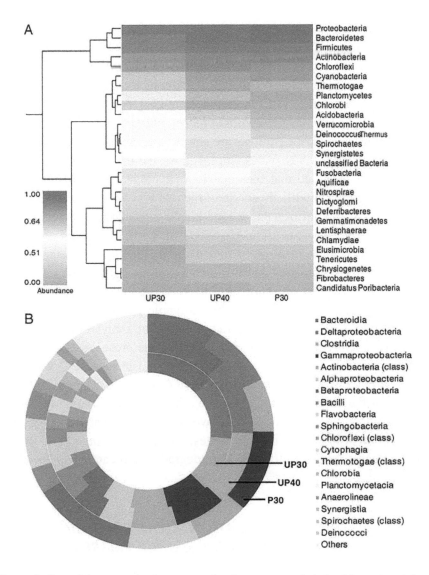

Figure 3: Bacterial community in pretreated and un-pretreated sludge bioreactor at day 30 and 40. (A) Proteobacteria, Bacteroidetes, Firmicutes, Actinobacteria and Chloroflexi represent top abundant phyla, all account for over 0.7 of the bacterial reads (normalized between 0 and 1) in the bioreactors. Abundance is displayed by both colour scheme and clustering dendrogram. (B) Bacterial consortiums which represent an anaerobic digestion community were manifested with overlapped dominant lineages in the bioreactors, with over-represented α-, β- and γ-proteobacteria in the sludge bioreactor consequent to the pretreatment (compare inner and outermost rings).

Acetoclastic and hydrogenotrohpic pathways are indeed common methanogenesis pathways reported in various reports [42,56]; the affirmation of these pathways permitted subsequent in-depth investigation of the encompassed enzymology. The first step in acetoclastic methanogenesis is the formation of acetyl-CoA from acetate [51]. Analysis of the recently completed genome of *Methanosaeta thermophilia* confirmed that the majority of the acetoclastic pathway are similar for *Methanosaeta* and *Methanosarcina*, except the enzymes employed for the first step of catalyzation [51]. It was proposed that acetoclastic methanogenesis in *Methanosaeta* proceeded with a modified version of the pathway compared with *Methanosarcina*, which utilizes the acetate kinase/phosphotransacetylase pathway to convert acetate to acetyl-CoA [57,58]. Taking *M. thermophila* as an example, its genome does not include a readily identifiable acetate kinase, and it has been postulated that this archeon utilizes an acetate transporter coupled with acetyl-CoA synthetases to convert acetate to acetyl-CoA, and the hydrolysis of pyrophosphate by inorganic pyrophosphatase (PPase) drives this reaction forward [51]. In this study, the analysis of metagenomic datasets using SEED annotation indicated the presence of acetyl-CoA synthetases (EC 6.2.1.1) and inorganic pyrophosphatase (EC 3.6.1.1) in the compiled *Methanosaeta* bin. While the sequencing depth might account for the absence of genes for the acetate kinase/phosphotransacetylase pathway, the discovery of these two enzymes advocated an alternative pathway in acetotrohpic methanogenesis for *Methanosaeta* [41]. Results from a recent study on a terephthalate-degrading bioreactor supported this hypothesis [41].

11.2.4 FUNCTIONAL AFFILIATION RELATED TO HIGHER METHANE PRODUCTION AFTER SLUDGE PRETREATMENT

Degradation of the highly organic polymers represents the first and overall rate-limiting step for the mineralization of organic matter in activated sludge and anaerobic digested sludge treatment systems [59-61]. In our work, the studied pretreatment involved alteration of pH, which is an important parameter affecting both bacterial activity and metabolite pathways. KO-based annotations were used to understand how these phylo-

Table 2: Relative abundance of Archaea in pretreated and un-pretreated sludge bioreactors at day 30 and 40

Archaeal taxonomic affiliation (phylum/family/genus/species)	Percentage of all Archaeal reads		
	UP30	UP40	P30
Euryarchaeota	97.60%	95.60%	94.86%
Methanosarcinaceae	23.11%	20.56%	19.14%
Methanosarcina	15.10%	13.32%	12.10%
Methanobacteriaceae	20.81%	18.61%	21.44%
Methanothermobacter	10.73%	9.41%	10.91%
Methanothermobacter thermautotrophicus	9.50%	8.59%	9.94%
Methanosaetaceae	14.00%	11.03%	9.19%
Methanosaeta	13.13%	10.31%	8.56%
Methanosaeta thermophila	14.48%	12.36%	10.25%
Crenarchaeota	2.33%	3.61%	4.40%
Korarchaeota	0.00%	0.45%	0.37%
Thaumarchaeota	0.05%	0.27%	0.37%
Nanoarchaeota	0.02%	0.07%	0.00%

Legends. Archaeal domain consists three major families belonging to Methanosarcinaceae, Methanobacteriaceae and Methanosaetaceae, while *Methanosarcina*, *Methanosaeta* and *Methanothermobacter* represent top abundant methanogens on the genus level. The decrease in abundance of *Methanosarcina*-related species in P30 was correlated with an enhanced bioproduction of acetate in the pretreated sludge. Except for phylum level, only candidates of abundance >8% are presented. Species are represented as percentage of all methanogens.

genetic trends could be used to predict the metabolic potential of these microbes. Figure 5 shows the subsystems that are related to higher methane production, including metabolism of amino acids, energy, carbohydrates, nucleotides, lipids, cofactors and vitamins, xenobiotics, as well as the fermentation of different substrates. These results revealed a general elevated expression of these faculties in the bioreactors fed with un-pretreated sludge over time (compare UP30 and UP40), whereas these levels were highest in the bioreactor digesting pretreated sludge (P30). Distribution of the functional systems was most divergent in P30, which showed predominance in metabolism consistent with a community shifted towards an enhanced biomass degradation metabolism. Herewith, the downstream analysis focused on the degradation of carbohydrates for a number of reasons. Firstly, the matrix of extra-cellular polymeric substances (EPS) that crosslink cells together remains the primary solid part of biological sludge, of which the polysaccharide is the predominant component [62-64]. Secondly, complex carbohydrate is the commonplace recalcitrant in the hydrolysis process, and bioprospecting for carbohydrate active enzymes is a topic of interest in metagenomics [65]. As a result, the focus of our study was hence placed on the degradation of this dominant recalcitrant component. (Please see Additional file 3 for further insights towards protein and lipid degradation.)

As sequencing technology provides access to a remarkable array of microbial functional capacity, sequence-based data mining is an important prospect in metagenomic projects [66,67]. To understand how the microbial community mediates the solibulization of the sludge matrix, a carbohydrate-active enzyme (CAZy) characterization of the metagenomic datasets was performed after ab initio gene prediction [68,69]. Conventional sequence homology-based enzyme discovery introduces a bias towards the identification of candidates similar to known enzymes, rather than enzymes with low sequence identity and potentially divergent biochemical properties [66]. The entries in the CAZy database contains both experimentally verified and putative carbohydrate-active enzyme domains, hence this search strategy would be able to provide a better insight into the catalysis of biochemical reactions [68,70].

In this study, a total of 1917 and 107 gene modules were recognized across 52 glycosyl hydrolases (GH) and 9 carbohydrate binding modules

(CBM) respectively (Tables 3 and 4). GH is defined as a widespread group of enzymes which hydrolyzes the glycosidic bond between carbohydrates or between that and a non-carbohydrate moiety [68]. On the other hand, CBM are contiguous amino acids within a carbohydrate-active enzyme with a discreet fold which bears the carbohydrate-binding activity [68]. In contrast to the small number of enzymes devoted to the hydrolysis of the main chain of cellulose, hemicelluloses and pectins (GH 5, 6, 7, 8, 9, 12, 44, 45, 48 and 74), all three metagenomes displayed a variety of enzymes that digest the side chains of these polymers and oligosaccharides. Families of GH 2 and 3, which contain a large range of glycosidases cleaving nonreducing carbohydrates in oligosaccharides and the side chains of hemicelluloses and pectins, were particularly abundant in the bioreactor samples (29.10% of all GH— UP30, 27.56%; UP40, 29.55%; P30, 24.20%). CBM 48 was the most dominant in the three metagenomes, it binds glycogen and is commonly found in a range of enzymes that act on branched substrates, such as the hydrolysis of glycogen, amylopectin and dextrin by isoamylase, pulluanase and branching enzyme. Overall, the number of GH hit returns appeared to increase with time in the bioreactor fed with un-pretreated sludge (468 counts in UP30; 792 counts in UP40); while the bioreactor digesting the pretreated sludge harboured 1.40 times the GH count in the control counterpart at the same time. As for CBM, P30 (51 counts) had a higher abundance than UP40 (38 counts), and made up to approximately 2.83 times that of UP30. Generally, there appeared to be a dynamic process in enzymology within the anaerobic digesters with time and substrate nature The heightened GH and CBM abundance in the bioreactor with pretreated sludge may possibly correlate with the higher sludge solubilization proposed in previous studies [23].

11.3 CONCLUSIONS

The treatment and disposal of excess sludge accumulation represents a bottleneck in wastewater treatment plants worldwide due to environmental, economic, social and legal factors. Reduction of excess sludge is becoming one of the biggest challenges in biological wastewater treatment [71,72]. The anaerobic digestion process is a promising technique, along

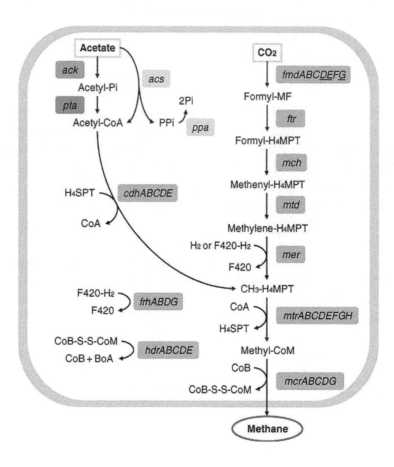

Figure 4: Reconstruction of methanogenesis pathways occurring in the bioreactors using identified genes. Positive identifications in meta-datasets are shown in colored boxes, with negative identifications shown underlined. Gene candidates for the formation of acetyl-CoA from acetate in the Methanosaeta bin are displayed in green boxes, and negative identifications particular to this bin are shown in purple boxes. See Additional file 2 for full form of abbreviated names and detailed identification counts.

with which waste is eliminated and methane is produced as a valuable renewable energy source. Understanding the impact of pretreatment on the
microbiology at different stages of anaerobic digestion is important; it ultimately impacts the performance of a bioreactor. This study represents the
first attempt to gain an in-depth metagenomic perspective, with regard to
taxonomic and functional aspects, on increased biogas (methane) production when a newly devised pretreatment method is used. Dual taxonomic
and functional analysis indicated the microbial and metabolic consortium
shifted extensively towards enhanced biodegradation, also seen in the
methanogenesis pathways. Altogether, these results presented here help
further microbiological understanding in sludge pretreatment research for
anaerobic digestion. As microbiology is the ultimate driver for the anaerobic digestion process, this new perspective encourages a closer engage-
ment between the engineering and microbiology knowledge pools. Further
studies involving deeper sequencing of the metagenomes to characterize
the decomposing and methanogenic microbiome at higher frequency and
replicate number parallel to the physical-chemical characterization are required for each WAS pretreatment intervention.

11.4 METHODS

11.4.1 COLLECTION OF WASTE ACTIVATED SLUDGE AND ANAEROBIC GRANULAR SLUDGE

The waste activated sludge (WAS) used as substrate for methane production was collected from the secondary sedimentation tank of a municipal
wastewater treatment plant in Shanghai, China. The anaerobic granular
sludge (AGS) used as the inoculum for methane production was obtained
from the upward-flow anaerobic sludge blanket (USAB) reactor of a food
wastewater treatment plant in Yixing, China. The main features of WAS
and AGS can be found in Additional file 4. The AGS was cultured in a
laboratory up-flow anaerobic sludge bed (UASB) by synthetic wastewater
(Additional file 5) prior to its use as the inoculum for methane production,
where the use of synthetic wastewater for AGS culturing is a common
practice in anaerobic digestion studies investigating the microbiology and

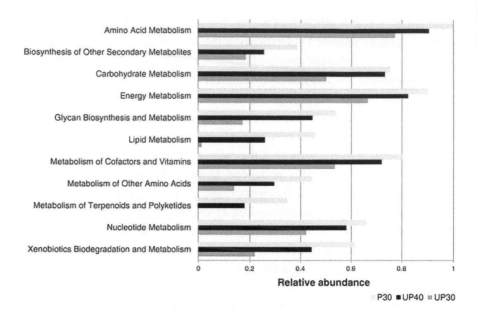

Figure 5: Relative abundance of functional reads affiliated to metabolisms associated to higher methane production using KO database. Expression of these faculties in the unpretreated sludge bioreactor gradually elevated over time (compare UP30 and UP40), and were highest in pretreated-sludge bioreactor (P30).

biodegradability [4,73,74]. The hydrolytic retention time of UASB was 6h, and the AGS concentration in UASB was approximately 29165 mg/L.

11.4.2 PRETREATMENT OF WASTE ACTIVATED SLUDGE

Biogas-producing, anaerobic digester samples were obtained from a re-production of the experiment described in our previous work [4], which introduced a new sludge pretreatment strategy (pH 10 for 8 d) for enhanced methane production from WAS. During the pretreatment, the pH of the collected WAS was adjusted to 10.0 by addition of 5 M NaOH or 4 M HCl with an automatic titrator, in a tightly-sealed reactor which was mechanically stirred (80 rpm) at $35\pm1°C$ for 8 d. A blank test was also conducted, in which the sludge was not pretreated but mechanically stirred for 8 d. Then the pH in all reactors was adjusted to pH 7.0, and 30 mL of AGS was added to each reactor for methane production. All reactors were sealed with rubber stoppers and mechanically stirred at 80 rpm without further pH control. The retention time of all reactors was 17 d and the organic loading rate was 1.88 kg/COD/m³/d. The gas generation was detected by water displacement and the methane concentration was measured by a gas chromatograph (GC-14B, Shimadzu, Japan). The analyses of VSS were conducted in accordance with APHA Standard Methods [75]. In this study, the methane yield was reported as the amount of methane generated per gram of volatile suspended solid added (mL-CH_4/gVSS- added) unless otherwise stated. The full unit of methane production is ml-CH_4/ gVSS-added, in which the VSS-added refers to the initial VSS to facilitate comparison of methane production. Moreover, there was no SCFA in the original sludge before anaerobic fermentation, meaning that the initial VSS did not contain volatile fatty acids—hence the risk of overestimation of specific methane yield is improbable [4].

11.4.3 PREPARATION OF ANAEROBIC DIGESTER SAMPLES

The granular samples cultured with pretreated (pH 10 for 8 d) and un-pretreated WAS were collected when the methane production rate reached the maximum on day 17, and then further cultured by a semi-continuous method to maintain the methane production rate. In replicates of three, the semi-continuous flow reactors with working volume of 1.0 L received 470 mL of pH 10 pretreated sludge and un-pretreated sludge, respectively. Everyday, 52 mL sludge mixture was manually wasted from each reactor, and 52 mL pretreated and un-pretreated sludge was added to the respective reactors; AGS was fed back to reactors by sifting the wasted sludge through a sifter with aperture of 0.2 mm. Further procedure details were described in [4].

As a stable methane generation continued to be maintained by the semi-continuous culturing method, 50 mL anaerobic digester samples were collected from the pretreatment set-ups (as pooled sample) at day 30 (P30), control set-up at day 30 (UP30) and control set-up at day 40 (UP40)

Table 3: Identification of glycoside hydrolase (GH) family classified by CAZy database

CAZy GH family	Pfam model	Number of counts			CAZy GH family	Pfam model	Number of counts		
		UP30	UP40	P30			UP30	UP40	P30
GH 1	PF00232.12	11	14	15	GH 38C	PF07748.7	20	42	15
GH 2	PF00703.15	7	27	28	GH 39	PF01229.11	2	1	6
GH 2_C	PF02836.11	18	38	31	GH 42	PF02449.9	8	18	12
GH 2_N	PF02837.12	26	61	44	GH 42C	PF08533.4	0	3	3
GH 3	PF00933.15	49	69	59	GH42M	PF08532.4	7	5	4
GH 3_C	PF01915.16	29	39	33	GH 43	PF04616.8	12	27	33
GH 4	PF02056.10	9	20	10	GH 44	PF12891.1	1	1	2
GH 8	PF01270.1	0	0	1	GH 4C	PF11975.1	8	18	8
GH 9	PF00759.1	9	5	7	GH 53	PF07745.7	1	1	10
GH 10	PF00331.14	4	10	11	GH 57	PF03065.1	22	38	12
GH 15	PF00723.1	2	2	7	GH 62	PF03664.7	0	1	0
GH 16	PF00722.15	11	9	9	GH 63	PF03200.10	1	1	0
GH 17	PF00332.12	0	0	1	GH 65C	PF03633.9	2	2	3

GH	PF				GH	PF			
GH 18	PF00704.22	7	11	7	GH65m	PF03632.9	19	5	12
G-I 19	PF00182.13	0	0	1	GH65N	PF03636.9	3	8	5
GH 20	PF00728.16	25	40	22	GH 67C	PF07477.6	2	5	1
GH 20b	PF02838.9	12	6	7	GH67M	PF07488.6	3	6	7
GH 25	PF01183.14	7	8	8	GH67N	PF03648.8	4	9	5
GH 26	PF02156.9	5	10	6	GH 76	PF03663.8	0	1	1
GH 28	PF00295.11	2	16	11	GH 77	PF02446.11	7	26	17
GH 30	PF02055.10	3	8	11	GH 81	PF03639.7	0	0	1
GH 31	PF01055.20	20	29	30	GH 88	PF07470.1	9	10	18
GH 32C	PF08244.6	0	5	0	GH 92	PF07971.6	34	49	46
GH32N	PF00251.14	4	4	3	GH 97	PF10566.3	15	49	26
GH 35	PF01301.13	12	4	18	GH 101	PF12905.1	7	3	7
GH 38	PF01074.16	8	26	21	GH cc	PF11790.2	1	2	2
					Total number of counts		468	792	657

Legend: Dynamic process in enzymology with time and substrate nature was observed in the bioreactors, the heightened GH and abundance (in bold) in the pretreated-sludge bioreactor were correlated with a higher sludge solubilization, and in turns a higher SCFA production.

for microbial studies. The two samples collected at day 30 (P30, UP30) served to characterize taxonomic and functional difference between pre-treated and unpretreated samples. Both sludge bioreactor were seeded with the same synthetic wastewater cultured AGS substrate, and later gradually replaced by the experimental sludge (pretreated or unpretreated WAS) completely as the batch experiment continued. Since both bioreactors received the same inoculums, therefore the only existing variable was the pretreatment, ascribing to the observed difference between the micro-biomes. As for UP40, this additional un-pretreated sample at day 40 was prepared to provide further insight of the microbial dynamic in response to time. Figure 1 shows the experimental scheme for the preparation of the anaerobic digester samples.

Table 4: Identification of carbohydrate binding module (CBM) family classified by CAZy database

CAZy CBM family	Pfam model	UP30	UP40	P30
CBM_2	PF00553.13	0	0	1
CBM_4_9	PF02018.11	1	5	6
CBM_5_12	PF02839.8	0	2	1
CBM_6	PF03422.9	2	6	8
CBM_11	PF03425.7	0	3	1
CBM_20	PF00686.13	0	4	3
CBM_25	PF03423.7	0	1	0
CBM_48	PF02922.12	11	10	20
CBM_X	PF06204.5	4	7	11
	Total number of counts	18	38	51

Legend: Dynamic process in enzymology with time and substrate nature was observed in the bioreactors, the heightened CBM abundance (in bold) in the pretreated-sludge bioreactor were correlated with a higher sludge solubilization, and in turns a higher SCFA production.

11.4.4 DNA EXTRACTION AND 454 SEQUENCING

Samples (P30, UP30 and UP40) were kept at −20 °C and thawed once only for nucleic acid extraction using PowerSoilW DNA Isolation Kit (Mobio, Carlsbad, USA), conducted within 24 hours of sample collection. DNA quantification was performed using Nanodrop 2000 spectrophotometer (Nanodrop Technologies, Wilmington, USA). Sequencing of total DNA derived from each was done by the shotgun sequencing approach on the 454 GS Junior system, and the sequencing quality was maintained by 454 propriety software. The raw sequence reads are submitted under the Sequence Read Archive (SRA) accession number SAMN01924662(P30), SAMN01924663 (UP30), SAMN01924664 (UP40).

11.4.5 METAGENOME ANALYSIS

To obtain a quantitative picture of the taxonomic composition, sequence datasets were characterized without a prior assembly step [56]. The unassembled sequence reads were submitted to the online metagenome analysis tool Metagenome Rapid Annotation using Subsystem Technology (MG-RAST analysis pipeline 3.0) [76]. Following the quality trimming with default setting, taxonomic and functional profiling was performed using the M5 non-redundant protein database (M5NR) and KEGG Orthology (KO) reference database [54] Parameters used were e-value cutoff of 1×10^{-5}, minimum alignment length of 50 base pairs, and a minimum percentage identity at 50%. Relative abundance of the Archaeal and Bacterial groups were determined by the percentages of respective reads over total assigned reads. Parallel processing was performed similarly on Megan4 to confirm the clustering pattern of taxonomic structure [77]. Only the taxonomic and functional annotation provided by MG-RAST was relied upon for all subsequent analysis.

To identify important carbohydrate-metabolism related gene candidates, pyrosequence reads were first subjected to gene prediction by a heuristic approach using MetaGeneMark 1.0 [69]. Next, the amino acid sequences were screened against catalytic domains (CD) characteristic for particular glycoside hydrolase (GH) families or for a particular carbohydrate binding module (CBM) as classified by the CAZy database [68]. HMM models were downloaded from Pfam database version 25.0 and used as a database for pfam_scan [78] (e-value cutoff of 1×10^{-4}). Methanogenesis related gene candidates were identified with reference to KEGG and Metacyc pathway database [54,55].

REFERENCES

1. Kim S, Kim J, Park C, Kim TH, Lee M, Kim SW, Lee J: Effects of various pretreatments for enhanced anaerobic digestion with waste activated sludge. J Biosci Bioeng 2003, 95:271–275.
2. Reynolds TDR, P A: Unit operations and process in environmental engineering. 2nd edition. Boston, MA, USA: PWS Publishing; 1996.
3. Jahng D, Cui R: Enhanced methane production from anaerobic digestion of disintegrated and deproteinized excess sludge. Biotechnol Lett 2006, 28:531–538.
4. Zhang D, Chen Y, Zhao Y, Zhu X: New sludge pretreatment method to improve methane production in waste activated sludge digestion. Environ Sci Technol 2010, 44:4802–4808.
5. Weemaes M, Grootaerd H, Simoens F, Verstraete W: Anaerobic digestion of ozonized biosolids. Water Res 2000, 34:2330–2336.
6. Shin SG, Lee S, Lee C, Hwang K, Hwang S: Qualitative and quantitative assessment of microbial community in batch anaerobic digestion of secondary sludge. Bioresource Technol 2010, 101:9461–9470.
7. Mossakowska A, Hellstrom BG, Hultman B: Strategies for sludge handling in the Stockholm region. Water Sci Technol 1998, 38:111–118.
8. Odegaard H, Paulsrud B, Karlsson I: Wastewater sludge as a resource: sludge disposal strategies and corresponding treatment technologies aimed at sustainable handling of wastewater sludge. Water Sci Technol 2002, 46:295–303.
9. Riviere D, Desvignes V, Pelletier E, Chaussonnerie S, Guermazi S, Weissenbach J, Li T, Camacho P, Sghir A: Towards the definition of a core of microorganisms involved in anaerobic digestion of sludge. ISME J 2009, 3:700–714.
10. Supaphol S, Jenkins SN, Intomo P, Waite IS, O'Donnell AG: Microbial community dynamics in mesophilic anaerobic co-digestion of mixed waste. Bioresource Technol 2011, 102:4021–4027.
11. Ting CH, Lin KR, Lee DJ, Tay JH: Production of hydrogen and methane from wastewater sludge using anaerobic fermentation. Water Sci Technol 2004, 50:223–228.
12. Raskin L, Griffin ME, McMahon KD, Mackie RI: Methanogenic population dynamics during start-up of anaerobic digesters treating municipal solid waste and biosolids. Biotechnol Bioeng 1998, 57:342–355.
13. Stams AJ: Metabolic interactions between anaerobic bacteria in methanogenic environments. Antonie Van Leeuwenhoek 1994, 66:271–294. 14.
14. Schink B: Syntrophic associations in methanogenic degradation. Prog Mol Subcell Biol 2006, 41:1–19.

15. Thauer RK: Biochemistry of methanogenesis: a tribute to Marjory Stephenson. 1998 Marjory Stephenson Prize Lecture. Microbiology 1998, 144(Pt 9):2377–2406.

16. Li YY, Noike T: Upgrading of anaerobic digestion of waste activated-sludge by thermal pretreatment. Water Sci Technol 1992, 26:857–866.

17. Vlyssides AG, Karlis PK: Thermal-alkaline solubilization of waste activated sludge as a pre-treatment stage for anaerobic digestion. Bioresource Technol 2004, 91:201–206.

18. Neis U, Tiehm A, Nickel K, Zellhorn M: Ultrasonic waste activated sludge disintegration for improving anaerobic stabilization. Water Res 2001, 35:2003–2009.

19. Tanaka S, Kobayashi T, Kamiyama K, Bildan MLNS: Effects of thermochemical pretreatment on the anaerobic digestion of waste activated sludge. Water Sci Technol 1997, 35:209–215.

20. Wei H, Tucker MP, Baker JO, Harris M, Luo Y, Xu Q, Himmel ME, Ding SY: Tracking dynamics of plant biomass composting by changes in substrate structure, microbial community, and enzyme activity. Biotechnol Biofuels 2012, 5:20.

21. Lau MW, Gunawan C, Dale BE: The impacts of pretreatment on the fermentability of pretreated lignocellulosic biomass: a comparative evaluation between ammonia fiber expansion and dilute acid pretreatment. Biotechnol Biofuels 2009, 2:30.

22. Yuan HY, Chen YG, Zhang HX, Jiang S, Zhou Q, Gu GW: Improved bioproduction of short-chain fatty acids (SCFAs) from excess sludge under alkaline conditions. Environ Sci Technol 2006, 40:2025–2029.

23. He PJ, Yu GH, Shao LM, He PP: Toward understanding the mechanism of improving the production of volatile fatty acids from activated sludge at pH 10.0. Water Res 2008, 42:4637–4644.

24. Venter JC, Remington K, Heidelberg JF, Halpern AL, Rusch D, Eisen JA, Wu DY, Paulsen I, Nelson KE, Nelson W, et al: Environmental genome shotgun sequencing of the Sargasso Sea. Science 2004, 304:66–74.

25. Sogin ML, Morrison HG, Huber JA, Mark Welch D, Huse SM, Neal PR, Arrieta JM, Herndl GJ: Microbial diversity in the deep sea and the underexplored "rare biosphere". P Natl Acad Sci USA 2006, 103:12115–12120.

26. Krober M, Bekel T, Diaz NN, Goesmann A, Jaenicke S, Krause L, Miller D, Runte KJ, Viehover P, Puhler A, Schluter A: Phylogenetic characterization of a biogas plant microbial community integrating clone library 16S-rDNA sequences and metagenome sequence data obtained by 454- pyrosequencing. J Biotechnol 2009, 142:38–49.

27. Hollister EB, Engledow AS, Hammett AJM, Provin TL, Wilkinson HH, Gentry TJ: Shifts in microbial community structure along an ecological gradient of hypersaline soils and sediments. ISME J 2010, 4:829–838.

28. Wirth R, Kovács E, Maróti G, Bagi Z, Rákhely G, Kovács KL: Characterization of a biogas-producing microbial community by short-read next generation DNA sequencing. Biotechnol Biofuels 2012, 5:4.

29. Zhang H, Banaszak JE, Parameswaran P, Alder J, Krajmalnik-Brown R, Rittmann BE: Focused-Pulsed sludge pre-treatment increases the bacterial diversity and relative abundance of acetoclastic methanogens in a full-scale anaerobic digester. Water Res 2009, 43:4517–4526.

30. Dinsdale EA, Edwards RA, Hall D, Angly F, Breitbart M, Brulc JM, Furlan M, Desnues C, Haynes M, Li LL, et al: Functional metagenomic profiling of nine biomes. Nature 2008, 452:629–U628.

31. Okabe S, Ariesyady HD, Ito T: Functional bacterial and archaeal community structures of major trophic groups in a full-scale anaerobic sludge digester. Water Res 2007, 41:1554–1568.

32. Tanner RS: Biology of Anaerobic Microorganisms - Zehnder, Ajb. Science 1989, 245:201–202.

33. Deppenmeier U: The unique biochemistry of methanogenesis. Prog Nucleic Acid Re 2002, 71:223–283.

34. Blackall LL, Burrell PC, O'Sullivan C, Song H, Clarke WP: Identification, detection, and spatial resolution of Clostridium populations responsible for cellulose degradation in a methanogenic landfill leachate bioreactor. Appl Environ Microb 2004, 70:2414–2419.

35. Klocke M, Mahnert P, Mundt K, Souidi K, Linke B: Microbial community analysis of a biogas-producing completely stirred tank reactor fed continuously with fodder beet silage as mono-substrate. Syst Appl Microbiol 2007, 30:139–151.

36. Sekiguchi Y, Takahashi H, Kamagata Y, Ohashi A, Harada H: In situ detection, isolation, and physiological properties of a thin filamentous microorganism abundant in methanogenic granular sludges: a novel isolate affiliated with a clone cluster, the green non-sulfur bacteria, subdivision I. Appl Environ Microb 2001, 67:5740–5749.

37. Blackall LL, Bjornsson L, Hugenholtz P, Tyson GW: Filamentous Chloroflexi (green non-sulfur bacteria) are abundant in wastewater treatment processes with biological nutrient removal. Microbiol-Sgm 2002, 148:2309–2318.

38. 38. Sekiguchi Y, Yamada T, Hanada S, Ohashi A, Harada H, Kamagata Y: Anaerolinea thermophila gen. nov., sp nov and Caldilinea aerophila gen. nov., sp nov., novel filamentous thermophiles that represent a previously uncultured lineage of the domain Bacteria at the subphylum level. Int J Syst Evol Micr 2003, 53:1843–1851.

39. Zakrzewski M, Goesmann A, Jaenicke S, Junemann S, Eikmeyer F, Szczepanowski R, Al-Soud WA, Sorensen S, Puhler A, Schluter A: Profiling of the metabolically active community from a production-scale biogas plant by means of high-throughput metatranscriptome sequencing. J Biotechnol 2012, 158:248–258.

40. Jaenicke S, Ander C, Bekel T, Bisdorf R, Droge M, Gartemann KH, Junemann S, Kaiser O, Krause L, Tille F, et al: Comparative and joint analysis of two metagenomic datasets from a biogas fermenter obtained by 454- pyrosequencing. PLoS One 2011, 6:e14519.

41. Liu WT, Lykidis A, Chen CL, Tringe SG, McHardy AC, Copeland A, Kyrpides NC, Hugenholtz P, Macarie H, Olmos A, Monroy O: Multiple syntrophic interactions in a terephthalate-degrading methanogenic consortium. ISME J 2011, 5:122–130.

42. Nelson MC, Morrison M, Yu Z: A meta-analysis of the microbial diversity observed in anaerobic digesters. Bioresour Technol 2011, 102:3730–3739.

43. Whitman WB, Liu YC: Metabolic, phylogenetic, and ecological diversity of the methanogenic archaea. Ann Ny Acad Sci 2008, 1125:171–189.

44. Raskin L, McMahon KD, Zheng DD, Stams AJM, Mackie RI: Microbial population dynamics during start-up and overload conditions of anaerobic digesters treating municipal solid waste and sewage sludge. Biotechnol Bioeng 2004, 87:823–834.

45. Raskin L, Stroot PG, McMahon KD, Mackie RI: Anaerobic codigestion of municipal solid waste and biosolids under various mixing conditions - I. Digester performance. Water Res 2001, 35:1804–1816

46. Liesack W, Grosskopf R, Janssen PH: Diversity and structure of the methanogenic community in anoxic rice paddy soil microcosms as examined by cultivation and direct 16S rRNA gene sequence retrieval. Appl Environ Microb 1998, 64:960–969.

47. Shigematsu T, Tang Y, Kobayashi T, Kawaguchi H, Morimura S, Kida K: Effect of dilution rate on metabolic pathway shift between aceticlastic and nonaceticlastic methanogenesis in chemostat cultivation. Appl Environ Microbiol 2004, 70:4048–4052.

48. Jetten MSM, Stams AJM, Zehnder AJB: Acetate threshold values and acetate activating enzymes in methanogenic bacteria. FEMS Microbiol Ecol 1990, 73:339–344.

49. Nettmann E, Bergmann I, Pramschufer S, Mundt K, Plogsties V, Herrmann C, Klocke M: Polyphasic analyses of methanogenic archaeal communities in agricultural biogas plants (vol 76, pg 2540, 2010). Appl Environ Microb 2011, 77:394–394.

50. Lee C, Kim J, Hwang K, O'Flaherty V, Hwang S: Quantitative analysis of methanogenic community dynamics in three anaerobic batch digesters treating different wastewaters. Water Res 2009, 43:157–165.

51. Smith KS, Ingram-Smith C: Methanosaeta, the forgotten methanogen? Trends Microbiol 2007, 15:150–155.

52. Smith DR, DoucetteStamm LA, Deloughery C, Lee HM, Dubois J, Aldredge T, Bashirzadeh R, Blakely D, Cook R, Gilbert K, et al: Complete genome sequence of *Methanobacterium thermoautotrophicum* Delta H: Functional analysis and comparative genomics. J Bacteriol 1997, 179:7135–7155.

53. Boone DW WB, Rouviere P: Diversity and taxonomy of methanogens. In Methanogenesis ecology, physiology, biochemistry & genetics. Edited by JG F. New York: Chapman & Hall; 1993:35–80.

54. Kanehisa M, Goto S: KEGG: Kyoto Encyclopedia of Genes and Genomes. Nucleic Acids Res 2000, 28:27–30.

55. Karp PD, Caspi R, Foerster H, Fulcher CA, Kaipa P, Krummenacker M, Latendresse M, Paley S, Rhee SY, Shearer AG, et al: The MetaCyc Database of metabolic pathways and enzymes and the BioCyc collection of Pathway/Genome Databases. Nucleic Acids Res 2008, 36:D623–D631.

56. Krause L, Diaz NN, Edwards RA, Gartemann KH, Kromeke H, Neuweger H, Puhler A, Runte KJ, Schluter A, Stoye J, et al: Taxonomic composition and gene content of a methane-producing microbial community isolated from a biogas reactor. J Biotechnol 2008, 136:91–101.

57. Metcalf WW, Rother M: Anaerobic growth of *Methanosarcina acetivorans* C2A on carbon monoxide: An unusual way of life for a methanogenic archaeon. P Natl Acad Sci USA 2004, 101:16929–16934.

58. Singhwissmann K, Ferry JG: Transcriptional regulation of the phosphotransacetylase-encoding and acetate kinase-encoding genes (Pta and Ack) from *methanosarcina-thermophila*. J Bacteriol 1995, 177:1699–1702.

59. Frolund B, Griebe T, Nielsen PH: Enzymatic-activity in the activated-sludge floc matrix. Appl Microbiol Biot 1995, 43:755–761.

60. Gessesse A, Dueholm T, Petersen SB, Nielsen PH: Lipase and protease extraction from activated sludge. Water Res 2003, 37:3652–3657.

61. Pletschke BI, Burgess JE: Hydrolytic enzymes in sewage sludge treatment: A mini-review. Water Sa 2008, 34:343–349.

62. Carrere H, Dumas C, Battimelli A, Batstone DJ, Delgenes JP, Steyer JP, Ferrer I: Pretreatment methods to improve sludge anaerobic degradability: a review. J Hazard Mater 2010, 183:1–15.

63. Cescutti P, Toffanin R, Pollesello P, Sutherland IW: Structural determination of the acidic exopolysaccharide produced by a *Pseudomonas* sp. strain 1.15. Carbohydr Res 1999, 315:159–168.

64. Sutherland IW, Kennedy L: Polysaccharide lyases from gellan-producing *Sphingomonas* spp. Microbiology 1996, 142(Pt 4):867–872.

65. Li LL, McCorkle SR, Monchy S, Taghavi S, van der Lelie D: Bioprospecting metagenomes: glycosyl hydrolases for converting biomass. Biotechnol Biofuels 2009, 2:10.

66. Rubin EM, Hess M, Sczyrba A, Egan R, Kim TW, Chokhawala H, Schroth G, Luo SJ, Clark DS, Chen F, et al: Metagenomic discovery of biomass-degrading genes and genomes from Cow rumen. Science 2011, 331:463–467.

67. Suen G, Scott JJ, Aylward FO, Adams SM, Tringe SG, Pinto-Tomas AA, Foster CE, Pauly M, Weimer PJ, Barry KW, et al: An insect herbivore microbiome with high plant biomass-degrading capacity. PLoS Genet 2010, 6(9): e1001129.

68. Cantarel BL, Coutinho PM, Rancurel C, Bernard T, Lombard V, Henrissat B: The Carbohydrate-Active EnZymes database (CAZy): an expert resource for Glycogenomics. Nucleic Acids Res 2009, 37:D233–238.

69. Zhu W, Lomsadze A, Borodovsky M: Ab initio gene identification in metagenomic sequences. Nucleic Acids Res 2010, 38:e132.

70. Li LL, Taghavi S, McCorkle SM, Zhang YB, Blewitt MG, Brunecky R, Adney WS, Himmel ME, Brumm P, Drinkwater C, et al: Bioprospecting metagenomics of decaying wood: mining for new glycoside hydrolases. Biotechnol Biofuels 2011, 4:23.

71. Chu LB, Wang JL, Wang B, Xing XH, Yan ST, Sun XL, Jurcik B: Changes in biomass activity and characteristics of activated sludge exposed to low ozone dose. Chemosphere 2009, 77:269–272.

72. Wei YS, Van Houten RT, Borger AR, Eikelboom DH, Fan YB: Minimization of excess sludge production for biological wastewater treatment. Water Res 2003, 37:4453–4467.

73. Bassin JP, Pronk M, Muyzer G, Kleerebezem R, Dezotti M, van Loosdrecht MCM: Effect of elevated salt concentrations on the aerobic granular sludge process: linking microbial activity with microbial community structure. Appl Environ Microb 2011, 77:7942–7953.

74. del Rio AV, Morales N, Isanta E, Mosquera-Corral A, Campos JL, Steyer JP, Carrere H: Thermal pre-treatment of aerobic granular sludge: Impact on anaerobic biodegradability. Water Res 2011, 45:6011–6020.

75. Eaton AD, Franson MAH, Association APH: Association AWW. Standard Methods for the Examination of Water & Wastewater. Federation WE: American Public Health Association; 1998.
76. Meyer F, Paarmann D, D'Souza M, Olson R, Glass EM, Kubal M, Paczian T, Rodriguez A, Stevens R, Wilke A, et al: The metagenomics RAST server - a public resource for the automatic phylogenetic and functional analysis of metagenomes. BMC Bioinforma 2008, 9:386.
77. Huson DH, Mitra S, Ruscheweyh HJ, Weber N, Schuster SC: Integrative analysis of environmental sequences using MEGAN4. Genome Res 2011, 21:1552–1560.
78. Eddy SR: Profile hidden markov models. Bioinformatics 1998, 14:755–763.

There are several supplemental files that are not available in this version of the article. To view this additional information, please use the citation on the first page of this chapter.

PART VI

BIOFILM BIOREACTORS

PART 6

BIOMATERIAL REACTORS

CHAPTER 12

Isolation and Molecular Characterization of Biofouling Bacteria and Profiling of Quorum Sensing Signal Molecules from Membrane Bioreactor Activated Sludge

HARSHAD LADE, DIBY PAUL AND JI HYANG KWEON

12.1 INTRODUCTION

Membrane bioreactors (MBR) have become state-of-the-art in wastewater treatment and have been in commercial use for more than two decades. The membrane filtration unit present in the MBR retains biomass and makes it effective for widespread use [1]. However, similar to other membrane separation technologies, membrane fouling still remains a major obstruction that limits its widespread use [2]. The occurrence of biofilm in MBR leads to membrane clogging and results in low treatment efficiency. Controlling biofilm formation in membranes can prevent MBR failures, reduce environmental damage and minimize revenue loss. In order to overcome the fouling problem, several operational strategies have been employed; however membrane fouling remains an issue of investigation in MBR operation.

Membrane biofouling is caused by several physicochemical and biological processes, which are highly dependent on composition of feedwater, membrane characteristics, operation conditions and microorganisms present [3]. The biofilm can be formed by a single bacterial species, but MBR biofilms almost always consist of mixed cultured communities from different genera. Therefore, it is essential to know the microbial community structure as a prerequisite for fundamental understanding of the biofouling phenomenon in MBR [4]. In addition, characterization of individual microbial genera will also give important information about bacteria involved in biofouling and therefore aid the design of control strategies. Culture-dependent techniques have been previously used for the isolation of a wide range of heterotrophic bacteria from activated sludge. Lim et al. [4] isolated a total of 61 bacterial species from membrane bioreactor biocake using culture-dependent techniques with various growth mediums. Specifically, they reported the predominance of *Enterobacter* and *Dyella* genera in MBR biocake bacterial isolates with a high number of *E. cancerogenus* strains. Biofouling bacteria typically present in activated sludge include the *Alpha-*, *Beta-* and *Gammaproteobacteria*, as well as the *Bacteroidetes* and the *Actinobacteria* [5]. The occurrence of different bacterial species from the genera *Aeromonas*, *Acinetobacter*, *Citrobacter*, *Klebsiella*, *Neisseria*, *Malikia* and *Pseudomonas* lineages have also been reported in activated sludge [6]. Miura et al. [3] found that *Betaproteobacteria* probably played the major role in development of mature biofilms in MBR treating municipal wastewater.

Biofilm formation begins with the colonization of bacteria and subsequent surface attachment. The initial colonization starts with the detection of quorum sensing (QS) signal molecules from the surrounding environment. The QS regulates the bacterial group behaviors in a cell density dependent way and is closely associated with biofouling of the membrane in MBR treating wastewater [7]. Thus characterization of activated sludge AHLs will provide an additional insight into the bacteria involved in MBR. Recently, Yeon et al. [7] reported three different AHLs: C6-HSL, C8-HSL and 3-oxo-C8-HSL in mixed cultured biocake of MBR treating wastewater. In addition, existence of C4-HSL and C6-HSL has been reported in soil isolate *Burkholderia* sp. strain A9 [8].

Developing more effective strategies to deal with the biofouling problem in MBRs treating wastewater still requires fundamental investigations on all aspects of biofilm formation. Several attempts have been made to know the physicochemical and biological parameters that lead to biofilm formation. Compared with physicochemical parameters, investigating the bacterial diversity responsible for MBR biofouling has been the main focus for future control strategies since these are the basic cause of membrane fouling. However, the microbial community responsible for biofouling is still not well characterized. To understand the bacteria responsible for biofouling and to choose the most appropriate control strategies, the isolation and molecular characterization of MBR activated sludge bacteria is essential. In addition, thorough exploration of the phylogenetic relationships within sludge bacteria and characterization of QS signal molecules will also provide additional insight into the actual fouling problem. However, such investigations do not allow for a reliable measure of microbial abundance and diversity of species, because the majority of the microbial population from the environment cannot be cultured [9]. But the approximate estimation and characterization of AHLs producing and biofouling bacterial diversity is essential for designing effective quorum sensing inhibitors (QSI) based control technologies.

The primary objectives of the present study were to characterize the key bacterial community responsible for membrane biofouling in a MBR treating wastewater, and profiling of QS signal molecules involved. In view of this, the culture-dependent isolation of activated sludge bacteria were carried out and initial screening for AHLs producing bacterial strains was performed with two wide range AHLs biosensor systems *C. violaceum* CV026 and *A. tumefaciens* A136. The correlation between AHLs production and biofilm formation among screened AHLs producing bacterial isolates has been made. The bacteria responsible for AHLs production and subsequent biofilm formation were identified by 16S rRNA gene sequence analysis and a phylogenetic relationship was established. The characterization of sludge AHLs was also performed by TLC and HPLC analysis, which will help us to design and investigate the QSI strategies for control of membrane biofouling.

12.2 RESULTS AND DISCUSSION

12.2.1 ISOLATION AND SCREENING OF AHLS PRODUCING BACTERIA

Isolation and characterization of AHLs producing bacteria becomes essential in the development of membrane biofouling control strategies. The culture dependent techniques allowed the isolation of a wide range of heterotrophic bacteria from environmental samples like activated sludge and marine sponges [4,10]. In the present study, a total of 200 morphologically distinct bacterial strains were randomly isolated from MBR activated sludge using culture-dependent techniques with NB, LB, TSB, R2A and SW medium respectively. Lim et al. [4] has isolated the QS active microbes from lab scale MBR using culture dependent techniques and correlated the microbial community with biofouling in MBR. Two different biosensor systems specific for short-chain AHLs (*C. violaceum* CV026) and long-chain AHLs (*A. tumefaciens* A136) were employed for screening AHLs producing bacterial strains. As a result, a total of 32 bacterial isolates were found to produce AHLs based on the development of purple and blue coloration in *C. violaceum* CV026 and *A. tumefaciens* A136 reporter strains, respectively. The well diffusion bioassay results of AHLs producing bacterial isolates are summarized in Table 1. The presence of 32 AHLs-producing and 168 non AHLs-producing strains in activated sludge sample isolates represent a model high-density microbial community of biotechnological importance. Among the 32 AHLs producers, 13 bacterial strains showed the ability to produce short or medium-chain AHLs with the induction of violacein production in reporter strain *C. violaceum* CV026 (Figure 1). However, all the 32 bacterial strains produce medium or long-chain AHLs as indicated by β-galactosidase activity with *A. tumefaciens* A136 (Figure 2). Similarly, Chong et al. [6] isolated and screened 52 morphologically distinct strains from activated sludge for production of QS signal molecules using various bioassays.

Table 1: Screening of bacterial isolates for AHLs production using violacein and β-galactosidase assay with *C. violaceum* CV026 and *A. tumefaciens* A136 reporter strains.

Isolate numbers	Isolate identifier	Reporter strains	
		C. violaceium CV026	*A. tumefaciens* A136
1	NA1	+++	+++
2	NA2	++	+++
3	NA3	--	++
4	NA4	++	+++
5	NA5	--	++
6	NA6	--	++
7	LBA1	+++	+++
8	LBA2	++	+++
9	LBA3	--	+++
10	LBA4	--	+++
11	LBA5	--	++
12	TSA1	+++	+++
13	TSA2	+++	+++
14	TSA3	--	+++
15	TSA4	--	++
16	TSA5	--	++
17	TSA6	--	++
18	TSA7	--	I+
19	TSA8	--	+++
20	R2A1	+++	+++
21	R2A2	+++	+++
22	R2A3	+++	++
23	R2A4	--	++
24	R2A5	--	++
25	R2A6	+++	+++
26	SWA1	++	+++
27	SWA2	--	+++
28	SWA3	--	++
29	SWA4	-	+++
30	SWA5	--	+++
31	SWA6	++	++
32	SWA7	--	++

++ Medium color production; +++ Strong color production.

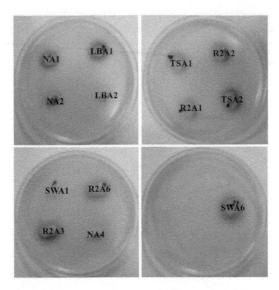

Figure 1: Screening of bacterial isolates for AHLs production using agar-plate well diffusion assay with *C. violaceum* CV026. All the thirteen isolates producing short and medium-chain AHLs are shown.

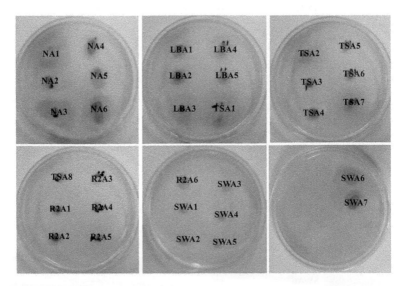

Figure 2: Screening of bacterial isolates for AHLs production using agar-plate well diffusion assay with *A. tumefaciens* A136. All the thirty two isolates producing medium and long-chain AHLs are shown.

12.2.2 BIOFILM FORMATION AMONG AHLS PRODUCING BACTERIAL ISOLATES

Most bacteria are social microorganisms that live in terrestrial environments and form highly organized and complex communities in the form of biofilms. Such biofilms develop on various surfaces including membrane bioreactor treating wastewater. It has been observed that all the 32 AHLs producing bacterial isolates form biofilms on polystyrene surfaces within 24 h incubation as assessed by the 96-well microplate assay. Quantification of biofilm produced by the 32 AHLs producing bacterial isolates revealed that all the strains seem to have different biofilm-forming abilities. Further prolonged incubation up to 48 h showed significant increases in biofilm formation by isolate no 6, 19, 21, 22, 23, 24 and 26 as compare to 24 h incubation (Figure 3). The ability of bacteria to establish biofilm can be used as a strategy for survival or for colonization on membrane surfaces [11,12]. Several recent studies have addressed that quorum sensing is the key factor which regulates biofilm formation by means of motility, adhesion, maturation and dispersal [13–16]. The present biofilm formation results are in good agreement with the previous studies which can establish the correlation between QS signal production and biofilm formation.

12.2.3 IDENTIFICATION OF AHLS PRODUCING BACTERIA

The accurate identification of bacterial isolates up to species level is crucial since this gives an insight into the bacterial diversity of the activated sludge. The identification of AHLs producing, as well as biofilm forming, bacterial isolates in this study was carried out by advanced molecular biology technique based on 16S rRNA gene sequence analysis. Analysis of nucleotide sequences has proved to be a powerful technique for phylogenetic characterization of bacteria [17]. The nucleotide blast analysis of complete 16S rRNA gene sequences form 32 bacterial isolates with the sequences available in the NCBI database demonstrated that they belong to eight different genera. The bacterial sequences comparison showed 98%–100% identification similarities with 16S rRNA gene sequences of the closest relative strains at NCBI database. The present molecular characterization reveals the significant differences in genetic diversity among AHLs producing bacterial strains. BLAST results of the 16S rRNA gene sequence analysis suggest that twelve isolates belong to genus *Aeromonas* (NA1, NA2, NA3, NA5, NA6, LBA1, LBA2, TSA1, TSA2, R2A1, R2A2, R2A3), ten to *Enterobacter* (NA4, LBA3, LBA4, LBA5, TSA5, R2A4, R2A6, SWA1, SWA3, SWA5), three each to *Serratia* (STSA8, SWA6, SWA7) and *Leclercia* (TSA4, TSA6, SWA4) while only one isolate was assigned each to *Pseudomonas* (TSA3), *Klebsiella* (SWA2), *Raoultella* (TSA7) and *Citrobacter* (R2A5) (Table 2). Molecular characterization study suggest the dominance of culturable *Aeromonas* and *Enterobacter* species, which are known to be common AHLs producing bacteria present in the membrane bioreactor activated sludge. Although most of the bacterial isolates belong to two dominant genera *Aeromonas* and *Enterobacter*, the molecular identification revealed some significant differences at the species level. Our results are in good agreement with those of Chong et al. who demonstrated distinct strains isolated from activated sludge have various genera and produce broad range of AHLs, which includes 17 *Aeromonas*, six *Acinetobacter*, five *Citrobacter*, four *Klebsiella*, two *Neisseria*, two *Malikia* and two *Pseudomonas* [6].

12.2.4 PHYLOGENETIC ANALYSIS

Based on assembled complete 16S rRNA gene sequence alignment, the phylogeny tree of 32 culturable AHLs producing bacterial isolates was constructed with the neighbor-joining method. This method demonstrates the position of each bacterial isolate in the phylogeny with bootstrap support. The result from the well-supported phylogeny with high resolution inner branches suggests the existence of 31 bacterial strains from *Enterobacteriaceae* family and only one from *Pseudomonadaceae* (Figure 4). Phylogenetic analysis also reveals that the isolate *Pseudomonas japonica* strain TSA3 forms a distinct clade and possibly belongs to a novel species with the *Pseudomonas japonica* strain IAM15071 as its closest relative. The *Aeromonas* and *Enterobacter* were dominant groups, which constituted 37% and 31% of the total AHLs producing isolates respectively. *Serratia* and *Leclercia* were the next group and constituted 9%, whereas

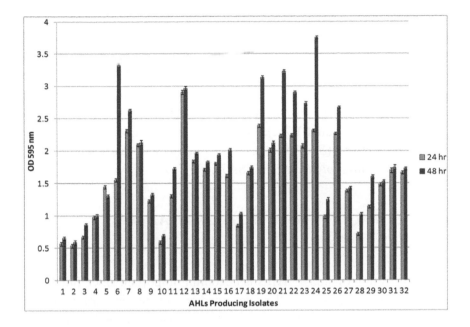

Figure 3: Biofilm formation among the AHLs producing bacteria as accessed by microtiter plate assay. The values are expressed as mean of four experiments and error bars indicates the standard deviation.

Pseudomonas, Klebsiella, Raoultella and *Citrobacter* constituted 3% each of total AHLs producing strains. Similar dominance by these groups was also reported in some AHLs producing bacterial isolation studies, where *Aeromonas* and *Enterobacter* species were found as dominant genera in bacterial quorum sensing signal producers [4,6,18]. Bacteria from the genus *Aeromonas, Pseudomonas, Enterobacter, Klebsiella, Serratia* and *Citrobacter* are well known in AHLs production and biofilm formation.

12.2.5 NUCLEOTIDE SEQUENCE ACCESSION NUMBERS

The eleven distinguishable AHLs producing bacterial isolates complete 16S rRNA gene sequences have been deposited in the GenBank database under the accession numbers: KF938658 (*Aeromonas hydrophila subsp. hydrophila* strain NA1), KF938659 (*Aeromonas media* strain NA2), KF938660 (*Aeromonas hydrophila subsp. dhakensis* strain LBA2), KF938661 (*Enterobacter ludwigi* strain SWA1), KF938662 (*Enterobacter* sp. strain LBA3), KF938663 (*Pseudomonas japonica* strain TSA3), KF938664 (*Enterobacter cancerogenus* strain LBA4), KF938665 (*Klebsiella variicola* strain SWA2), KF938666 (*Citrobacter freundii* strain R2A5), KF938667 (*Serratia marcescens* strain SWA6) and KF938668 (*Raoultella ornithinolytica* strain TSA7).

12.2.6 EXTRACTION AND CHARACTERIZATION OF SLUDGE AHLS

Characterization of sludge AHLs is becoming important since future biofouling control strategies will be designed by considering QS inhibition. In the present study, AHLs extracted from MBR activated sludge with acidified ethyl acetate were initially analyzed for the presence of a broad spectrum of AHLs with two biosensor strains. The cross-feeding agar plate assay with *C. violaceum* CV026 and *A. tumefaciens* A136 confirms the presence of a broad-range of AHLs in activated sludge extract (Figure 5). This method has the advantage of rapid detection of AHLs with approximate types.

Separation of AHLs by TLC bioassay with reporter strains gives an identifying index of AHLs present in the sludge extract. The TLC chromatogram overlayed with biosensor strains *C. violaceum* CV026 showed three well differentiated spots indicating the presence of short-chain AHLs in activated sludge extracts. Relative retention factor (Rf) of developed purple colour spots were calculated and compared with those of standard AHLs. The crude sludge AHLs extracts showed close resemblance with standard C4-HSL (Rf = 0.72), C6-HSL (Rf = 0.43) and C8-HSL (Rf = 0.21) (Figure 6). Further TLC analysis for long-chain AHLs by β-galactosidase assay with *A. tumefaciens* A136 showed a tailing pattern for sludge extracted AHLs and there was no clear separation of spots (data not shown).

HPLC analysis was performed to identify sludge AHLs by comparing appeared peaks retention time with those of respective standard AHLs.

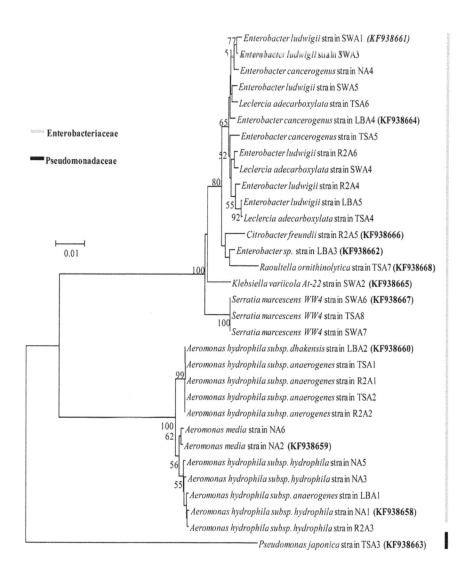

Figure 4: Phylogenetic relationship of the AHLs producing bacteria isolated from membrane bioreactor activated sludge. The tree was constructed using neighbor joining algorithm with Kimura 2 parameter distances in MEGA 4.0 software. Numbers at the nodes indicate percent bootstrap values above 50 supported by 550 replicates. The bar indicates the Juke-Cantor evolutionary distance.

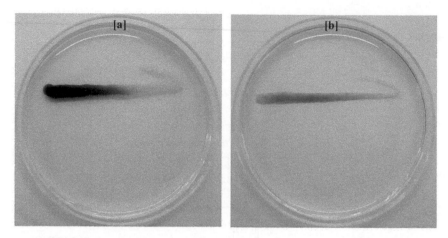

Figure 5: Detection of AHLs extracted from activated sludge by agar-plate cross feeding violacein and β-galactosidase assay with *C. violaceum* CV026 and *A. tumefaciens* A136 respectively. (a) Evidence for the presence of short and medium-chain AHLs in activated sludge extract is indicated by purple coloration of the reporter strain *C. violaceum* CV026; (b) Evidence for the presence of medium and long-chain AHLs activated sludge extract is indicated by blue coloration of the reporter strain *A. tumefaciens* A136.

The standard AHLs showed retention times of 3.221 for C4-HSL, 6.247 for C6-HSL, 11.932 for 3-oxo-C8-HSL, 16.168 for C8-HSL, 23.106 for C10-HSL, 25.209 for 3-oxo-C12-HSL, 28.165 for C12-HSL and 33.284 for C14-HSL (Figure 7a). The HPLC chromatogram of sludge extract showed a total of 13 peaks at various retention times. Eight peaks appeared with the retention times of 3.224 for C4-HSL, 6.296 for C6-HSL, 11.826 for 3-oxo-C8-HSL, 16.012 for C8-HSL, 23.121 for C10-HSL, 25.256 for 3-oxo-C12-HSL, 28.103 for C12-HSL and 33.212 for C14-HSL are identical to standard AHLs (Figure 7b). Furthermore, appearance of minor peaks at retention times of 10.354 and 14.445 min and the major peaks at 36.324 and 38.123 min that do not resemble standard AHLs suggest the presence of unidentified AHLs or microbial metabolites in the sludge extract. The appearance of the first major peak at retention time of

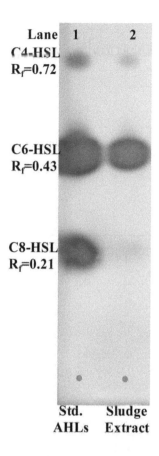

Figure 6: TLC bioassay with *C. violaceum* CV026 for the identification of AHLs extracted from activated sludge. Lane 1: Standard AHLs (0.5 μg/μL); C4-HSL, C6-HSL and C8-HSL. Lane 2: Extract of the activated sludge (20 μL). Tentative identification of the sludge extracted AHLs based on migration of standards, is indicated.

2.112 was for solvent. From the HPLC analyses it can be concluded that the presence of eight broad spectrum AHLs in the activated sludge sample may be due to the active involvement of AHLs producing bacteria from the family *Enterobacteriaceae* and *Pseudomonadaceae* as confirmed by 16S rRNA gene sequence analysis and phylogenetic relationship.

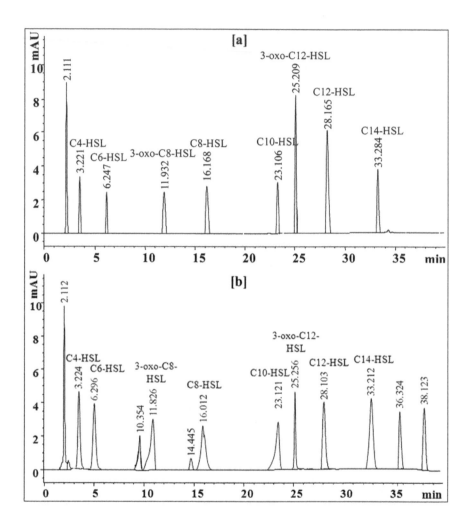

Figure 7: Identification of the AHLs extracted from MBR activated sludge by HPLC. (a) Chromatogram showing the retention times of standard AHLs each at 50 μg/mL concentration in a mixture of C4-HSL, C6-HSL, C8-HSL, 3-oxo-C8-HSL, C10-HSL, C12-HSL, 3-oxo-C12-HSL and C14-HSL; (b) The sludge AHLs were identified by comparing appeared peaks retention time with those of respective standard AHLs.

12.3 EXPERIMENTAL SECTION

12.3.1 CHEMICALS AND MICROBIOLOGICAL MEDIA

HPLC grade ethyl acetate, methanol, acetonitrile, water and dimethyl-formamide were purchased from Fisher Scientific (Fairlawn, NJ, USA). 5-bromo-4-chloro-3-indolyl-β-D-galactopyranoside (X-gal), kanamycin sulfate, spectinomycin and tetracycline were obtained from Sigma-Aldrich (St. Louis, MO, USA). Dehydrated culture media used includes Luria-Bertani (LB), Nutrient Broth (NB), Tryptic Soy Broth (TSB) and Reasoner's 2A (R2A) broth were procured from BD-Difco (Franklin Lakes, NJ, USA). All the chemicals used were of highest analytical grade.

12.3.2 STANDARD N-ACYL HOMOSERINE LACTONES

Standard AHLs viz. *N*-butanoyl-L-homoserine lactone (C4-HSL), *N*-hexanoyl-L-homoserine lactone (C6-HSL), *N*-octanoyl-L-homoserine lactone (C8-HSL), 3-oxo-octanoyl-L-homoserine lactone (3-oxo-C8-HSL) and *N*-decanoyl-L-homoserine (C10-HSL) were purchased from Sigma-Aldrich (St. Louis, MO, USA). The other long-chain AHLs *N*-dodecanoyl-L-homoserine (C12-HSL), *N*-(3-oxo-dodecanoyl)-L-homoserine lactone (3-oxo-C12-HSL) and *N*-tetradecanoyl-L-homerisine lactone (C14-HSL) were purchased from the Cayman Chemical Company (Ann Arbor, MI, USA).

12.3.3 QUORUM SENSING REPORTER'S STRAINS

Two different AHL biosensor systems *C. violaceum* CV026 (Mini-Tn5 mutant of ATCC31532) and *A. tumefaciens* A136 which are deficient in AHLs production were used to screen the wide range of AHLs producing bacterial isolates. These biosensor strains detect and respond to a range of exogenous AHLs molecules with an acyl side chain of four to fourteen carbons by inducing the synthesis of the purple pigment violacein and blue coloration by β-galactosidase activity. *C. violaceum* 026 bears a LuXR homologue, CviR, which regulates the production of violacein, a purple pigment when induced by C4-HSL, C6-HSL, C8-HSL and 3-oxo-C4~C8 exogenous AHLs [19,20]. *A. tumefaciens* A136 bears the traI-lacZ fusion (pCF218) (pCF372) plasmids and capable of producing a blue colour from the hydrolysis of 5-bromo-4-chloro-3-indolyl-β-D-galactopyranoside by the β-galactosidase activity, in response to C8-HSL, 3-oxo-C8-HSL, C10-HSL, C12-HSL, 3-oxo-C12-HSL and C14-HSL exogenous AHLs molecule [7,20,21]. Stock cultures of the reporter strains were stored at −80 °C in Luria-Bertani broth with 20% glycerol. The biosensor strains were aerobically activated in LB broth with shaking at 28 °C for 24 h. When required, LB broth was supplemented with the appropriate amount of antibiotics to maintain the plasmids (*C. violaceum* CV026 with kanamycin 20 μg/mL; *A. tumefaciens* A136 with spectinomycin 50 μg/mL and tetracycline 4.5 μg/mL). The pH of the culture medium was adjusted to 7.0 ± 0.2 with 0.1 N solution of NaOH or HCL.

12.3.4 COLLECTION OF ACTIVATED SLUDGE

Activated sludge samples were collected in sterile umber color bottles from the MBR of GURI wastewater treatment plant, Guri-city, Gyeonggi-do, Republic of Korea in summer 2013. The collected sludge was transported to laboratory in ice box and extraction was carried out immediately. The pH of the MBR activated sludge samples was 6.74.

12.3.5 ISOLATION OF BACTERIA

Using a culture-dependent technique with various growth media, the isolation of bacteria from membrane bioreactor activated sludge was performed [4]. Different microbiological culture medium used in this study were LB broth (g L^{-1} of tryptone, 10.0; yeast extract, 5.0; sodium chloride, 10.0), NB (g L^{-1} of peptic digest of animal tissue, 5.0; sodium chloride, 5.0; beef extract, 1.5; yeast extract, 1.5), TSB (g L^{-1} of pancreatic digest of casein,

17.0; enzymatic digest of soybean meal, 3.0; dextrose, 2.5; sodium chloride, 5.0; dipotassium phosphate, 2.5), R2A broth (g L^{-1} of yeast extract, 0.5; protease peptone No.3, 0.5; casamino acids, 0.5; dextrose, 0.5; soluble starch, 0.5; sodium pyruvate, 0.3, dipotassium phosphate, 0.3; magnesium sulfate, 0.05) and synthetic wastewater (SW) (g L^{-1} of glucose, 1.0; yeast extrxct, 0.05; bactopeptone, 0.05; ammonium sulfate, 0.5; dipotassium hydrogen phosphate, 0.3; potassium dihydrogen phosphate, 0.3; magnesium sulphate heptahydrate, 0.009; iron (III) chloride hexahydrate, 0.0002; sodium chloride, 0.007; calcium chloride, 0.0002; sodium hydrogen carbonate 0.15). The activated sludge sample was diluted up to 10^{-6} with sterile distilled water and 100 µL was spread inoculated onto each growth medium agar plates. Such plates were then incubated at 28 °C for 24 to 48 h and several morphologically distinct pure colonies were randomly chosen, subcultured on the same agar and grown in the same conditions. A total of 200 bacterial isolates, 40 from each medium agar plate, were selected and screened for AHLs production using two different biosensor systems. The stock cultures of bacterial isolates were maintained on LB agar at 4 °C.

12.3.6 SCREENING OF BACTERIAL ISOLATES FOR AHLS PRODUCTION

As different biosensor systems detect specific types of AHLs, it is essential to use those biosensors that respond to a broad range of AHLs. Screening of all the bacterial isolates for production AHLs was carried out by agar plate diffusion assay with some modifications [19,22]. Briefly, active cultures of QS reporter bacterial strains were prepared in LB agar plates containing the antibiotics kanamycin 20 µg/mL for *C. violaceum* CV026 and spectinomycin 50 µg/mL and tetracycline 4.5 µg/mL for *A. tumefaciens* A136. The plates were incubated at 28 °C for 24 h. A loopful of overnight grown cultures from both the plates were taken and separately added into 10 mL of sterile distilled water. Then 2.5 mL culture suspension of both biosensor strains were separately added to 100 mL of cooled LB medium (0.8% agar) kept at 45 °C and poured as a thin plate. 40 µg/mL of X-gal as a visualizing agent was incorporated into LB medium used for *A. tumefaciens* A136. The bacterial isolates to be tested were grown in LB broth for 24 h and diluted to an OD 600 nm of 1.0 with sterile distilled water. About 5 μL of the test strains were then spot inoculated onto the surface of previously prepared chromoplates and allowed to dry. The petri plates were incubated for 24 to 48 h at 28 °C and production of purple pigment violacein and blue coloration by β-galactosidase activity was observed as positive test for AHLs production.

12.3.7 MICROTITER PLATE BIOFILM FORMATION ASSAY

AHLs producing bacterial isolates were evaluated for biofilm formation in 96-well flat bottom polystyrene microtiter plate (SPL Life Sciences, Pocheon-Si, Korea) by the method of O'Toole and Kolter [23] with slight modifications. The conditioning of microplate wells was carried out by inoculating 200 μL of LB broth for 1 h at room temperature. The wells were emptied and 190 μL of LB medium and 10 μL of bacterial cell suspensions (0.5 O.D $_{.595nm}$) were inoculated in each well. The plate was sealed with parafilm and incubated in static condition at 28 °C for 24 and 48 h [24,25]. Following static incubation, the turbidity of the planktonic cultures was measured at 655 nm and the biofilm formation was quantified by classical crystal violet assay. Briefly, the supernatant from the microtiter plate wells was gently removed and the wells were rinsed thrice with 200 μL of sterile distilled water. The plate was dried in inverted position for 30 min and stained with 200 μL of 0.1% (*w/v*) crystal violet, per well, for 30 min. Excess crystal violet was removed from the wells and the biofilms were washed twice with distilled water. Finally, the absorbed crystal violet was extracted with 200 μL of 95% ethanol, per well for 1 h, and the O.D $_{.595nm}$ was measured with a plate reader (iMark™ Micrplate Absorbance Reader 168-1135, Bio-Rad, Hercules, CA, USA), which yields a measure of biofilm formation. A well with sterile LB broth served as controls.

12.3.8 IDENTIFICATION OF AHLS PRODUCING BACTERIA

The screened AHLs producing bacterial isolates were identified through 16S rRNA gene sequences analysis. The bacterial genomic DNA was extracted using the GenElute™ Bacterial genomic DNA Kit (Sigma-Aldrich, St. Louis, MO, USA), according to the manufacturer's instructions. 16S rRNA gene sequences were amplified with bacterial universal primers 518F (5'-CCAGCAGCCGCGGTAATACG-3') and the reverse primer 800R (5'-TACCAGGGTATCTAATCC-3'). The purified polymerase chain reaction (PCR) amplicons were then sequenced and compared with the 16S rRNA gene sequences available in the GeneBank database at the National Centre for Biotechnology Information (NCBI) using the basic local alignment search tool BLASTN 2.2.28 algorithm [26].

12.3.9 PHYLOGENETIC ANALYSIS OF AHLS PRODUCING BACTERIA STRAINS

All the 32 AHLs producing bacterial isolates nucleotide sequences were aligned with the program CLUSTAL W [27]. The aligned complete 16S rRNA sequences were subjected to phylogenetic analysis using the Molecular Evolutionary Genetic Analysis (MEGA) software version 4.0. The tree was generated with the Neighbor-Joining algorithm and bootstrap for 550 resamplings to ensure robustness and reliability of trees constructed [28].

12.3.10 NUCLEOTIDE SEQUENCE ACCESSION NUMBER

The assembled complete 16S rRNA sequences of eleven distinguishable AHLs producing bacterial isolates have been deposited in public sequence repository NCBI GenBank using the BankIt sequence submission tool (http://www.ncbi.nlm.nih.gov/Genbank; National Institutes of Health, Rockville, MD, USA).

12.3.11 EXTRACTION OF AHLS FROM ACTIVATED SLUDGE

The crude extraction of AHLs from the MBR activated sludge from GURI wastewater treatment plant was carried out with acidified ethyl acetate [29]. Briefly, 20 mL of activated sludge was centrifuged at 9000 g for

10 min and the supernatant was mixed with an equal volume of acidified ethyl acetate (0.1% acetic acid). The mixture was then shaken at 180 rpm for 2 h at 25 °C. After shaking, the upper organic layer was separated and freeze-dried under vacuum. The residue was then dissolved in 250 μL of acetonitrile and used further for chromatographic analysis.

12.3.12. DETECTION OF CRUDE AHLS BY AGAR PLATE CROSS-FEEDING BIOASSAY

In order to detect the occurrence of AHLs in activated sludge extract, the samples were analyzed by agar plate cross-feeding assay with two biosensor systems [21]. The LB agar plates containing kanamycin 20 μg/mL for *C. violaceum* CV026 and spectinomycin 50 μg/mL, tetracycline 4.5 μg/mL and X-gal 80 μg/mL for *A. tumefaciens* A136 were prepared. A loopful of crude AHLs extract was streak inoculated in parallel with approximately 1 cm to the *C. violaceum* CV026 and *A. tumefaciens* A136 biosensor strains on respective agar plates. After incubation for overnight at 28 °C the plates were observed for activation of LuXR homologue CviR and traI-lacZ fusion in *C. violaceum* CV026 and *A. tumefaciens* A136 reporter strains respectively to confirm the presence of AHLs. The occurrence of AHLs in crude extract is indicated by the development of purple and blue coloration of *C. violaceum* CV026 and *A. tumefaciens* A136 streaked lines respectively.

12.3.13 THIN LAYER CHROMATOGRAPHY (TLC) FOR AHL IDENTIFICATION

Ten microlitre of crude sludge AHLs preparation was spotted on a C18 reverse-phase TLC plate (60RP-18F$_{254}$S, Merck, Germany) and the chromatogram was developed in methanol: water (60:40) solvent system until the solvent front line reaches up to 1 cm from top edge [7, 29]. After being completely dried in clean bench for 20 min, the TLC plate was separately overlaid with 1.5% LB agar containing *C. violaceum* CV026 (kanamycin 20 µg/mL) and *A. tumefaciens* A136 (spectinomycin 50 µg/mL, tetracycline 4.5 µg/mL and X-gal 80 µg/mL). The commercially available AHLs prepared in acetonitrile were used as reference standards. TLC plate loaded with crude AHLs extract and biosensor strains was incubated overnight at 28 °C and observed for the apparent of purple and blue spots respectively. The results were captured digitally and the Rf value of standard AHLs, defined by the ratio of the distance traveled by spot and that of solvent front were calculated. The approximate type of extracted sludge AHLs were determined by comparing the Rf values with those of standard reference AHLs.

12.3.14. HIGH PERFORMANCE LIQUID CHROMATOGRAPHY (HPLC) ANALYSIS OF AHLS

Analysis of AHLs extract was performed on an Agilent 1200 HPLC system (Agilent, Santa Clara, CA, USA) equipped with a ZORBAX Eclipse XDB-C18 column (4.6 mm × 150 mm, 5 µm particle size) kept at 30 °C. Commercially available standard AHLs viz. C4-HSL, C6-HSL, C8-HSL, 3-oxo-C8-HSL, C10-HSL, C12-HSL, 3-oxo-C12-HSL and C14-HSL were dissolved in acetonitrile to make 1000 µg/mL of stock solutions. Each of the eight AHLs were further diluted in acetonitrile to obtain 10, 20, 30, 40 and 50 µg/mL of individual and its mixtures working solutions. The resulting AHLs standards, its mixture, crude sludge extract and acetonitrile as blank were injected into a column at a flow rate of 0.25 mL/min. The HPLC conditions and instrument parameters as previously described includes an isocratic profile of methanol/water (35:65, *v/v*) for five min, followed by a linear gradient from 35% to 95% methanol in water over 33 min. A subsequent linear gradient from 95% to 35% methanol in water over 2 min, and an isocratic profile of methanol/water (35:65, *v/v*) for five min were applied for flushing the column for the following run [21,30].

12.3.15. STATISTICAL ANALYSIS

Statistical analysis of biofilm formation was performed using SPSS software (IBM Corporation, Armonk, NY, USA). The values presented are means of four repeated experiments with standard deviations indicated in error bars.

12.4 CONCLUSIONS

This study represents the culture-dependent molecular approach to elucidate the molecular diversity and phylogenetic relationship of AHLs producing bacterial communities in a MBR treating wastewater. Our data have shown that AHLs producing bacteria were distinct from each other and reflect the presence of various genera. 16S rRNA gene sequence characterization indicates the predominance and thus the potential role of Aeromonas and Enterobacter species in AHLs production and subsequent biofilm formation. Furthermore, evidence for the presence of broad range of AHLs in activated sludge extract and the ability of respective bacterial isolates to produce such AHLs suggest that these genera could be main targets for fouling control in MBR. This study is an important step towards exploring the future QSI strategies to control biofouling in MBR treating wastewater.

REFERENCES

1.   Drews, A. Membrane fouling in membrane bioreactors—Characterisation, contradictions, cause and cures. J. Membr. Sci. 2010, 363, 1–28.
2.   Lee, J.W.; Jutidamrongphan, W.; Park, K.Y.; Moon, S.; Park, C. Advanced treatment of wastewater from food waste disposer in modified Ludzack-Ettinger type membrane bioreactor. Environ. Eng. Res. 2012, 17, 59–63.
3.   Miura, Y.; Watanabe, Y.; Okabe, S. Membrane biofouling in pilotscale membrane bioreactors (MBRs) treating municipal wastewater: Impact of biofilm formation. Environ. Sci. Technol. 2007, 41, 632–638.
4.   Lim, S.Y.; Kim, S.; Yeon, K.M.; Sang, B.I.; Chun, J.; Lee, C.H. Correlation between microbial community structure and biofouling in a laboratory scale membrane bioreactor with synthetic wastewater. Desalination 2012, 287, 209–215.

5. Wagner, M.; Loy, A. Bacterial community composition and function in sewage treatment systems. Curr. Opin. Microbiol. 2002, 13, 218–227.
6. Chong, G.; Kimyon, O.; Rice, S.A,; Kjelleberg, S.; Manefield, M. The presence and role of bacterial quorum sensing in activated sludge. Microb. Biotechnol. 2012, 5, 621–633.
7. Yeon, K.M.; Cheong, W.S.; Oh, H.S.; Lee, W.N.; Hwang, B.K.; Lee, C.H.; Beyenal, H.; Lewandowski, Z. Quorum sensing: A new biofouling control paradigm in a membrane bioreactor for advanced wastewater treatment. Environ. Sci. Technol. 2009, 43, 380–385.
8. Chen, J.W.; Koh, C.L.; Sam, C.K.; Yin, W.F.; Chan, K.G. Short chain N-acyl homoserine lactone production by soil isolate Burkholderia sp. strain A9. Sensors 2013, 13, 13217–13227.
9. Hugenholtz, P.; Goebel, B.M.; Pace, N.R. Impact of culture-independent studies on the emerging phylogenetic view of bacterial diversity. J. Bacteriol. 1998, 180, 4765–4774.
10. Alex, A.; Silva, V .; V asconcelos, V .; Antunes, A. Evidence of unique and generalist microbes in distantly related sympatric intertidal marine sponges (Porifera: Demospongiae). PLoS One 2013, 8, e80653.
11. Johnson, L.R. Microcolony and biofilm formation as a survival strategy for bacteria. J. Theor. Biol. 2008, 251, 24–34.
12. Rinaudi, L.V.; Giordano, W. An integrated view of biofilm formation in rhizobia. FEMS Microbiol. Lett. 2010, 304, 1–11.
13. Dong, Y.H.; Zhang, X.F.; An, S.W.; Xu, J.L.; Zhang, L.H. A novel two-component system BqsS-BqsR modulates quorum sensing-dependent biofilm decay in Pseudomonas aeruginosa. Commun. Integr. Biol. 2008, 1, 88–96.
14. Rasmussen, T.B.; Bjarnsholt, T.; Skindersoe, M.E.; Hentzer, M.; Kristoffersen, P.; Kote, M.; Nielsen, J.; Eberl, L.; Givskov, M. Screening for quorum-sensing inhibitors (QSI) by use of a novel genetic system, the QSI selector. J. Bacteriol. 2005, 187, 1799–1814.
15. Daniels, R.; Vanderleyden, J.; Michiels, J. Quorum sensing and swarming migration in bacteria. FEMS Microbiol. Rev. 2004, 28, 261–289.
16. Stoodley, P.; Wilson, S.; Hall-Stoodley, L.; Boyle, J.D.; Lappin-Scott, H.M.; Costerton, J.W. Growth and detachment of cell clusters from mature mixed-species biofilms. Appl. Environ. Microbiol. 2001, 67, 5608–5613.
17. 17. Adegboye, M.F.; Babalola, O.O. Phylogenetic characterization of culturable antibiotic producing Streptomyces from rhizospheric soils. Mol. Biol. 2013, S1, doi:10.4172/2168-9547.S1-001.
18. Wang, H.; Cai, T.; Weng, M.; Zhou, J.; Cao, H.; Zhong, Z.; Zhu, J. Conditional production of acyl-homoserine lactone-type quorum-sensing signals in clinical isolates of enterobacteria. J. Med. Microbiol. 2006, 55, 1751–1753.
19. McClean, K.H.; Winson, M.K.; Fish, L.; Taylor, A.; Chhabra, S.R. Quorum sensing and Chromobacterium violaceum: Exploitation of violacein production and inhibition for the detection of N-acylhomoserine lactones. Microbiology 1997, 143, 3703–3711.
20. Ravn, L.; Christensen, A.B.; Molin, S.; Givskov, M.; Gram, L. Methods for detecting acylated homoserine lactones produced by Gram-negative bacteria and their ap-

plication in studies of AHL-production kinetics. J. Microbiol. Methods 2001, 44, 239–251.

21. Kim, S.R.; Oh, H.S.; Jo, S.J.; Yeon, K.M.; Lee, C.H.; Lim, D.J.; Lee, C.H.; Lee, J.K. Biofouling control with bead-entrapped quorum quenching bacteria in membrane bioreactors: Physical and biological effects. Environ. Sci. Technol. 2013, 47, 836–842.

22. Anbazhagan, D.; Mansor, M.; Yan, G.O.S.; Yusof, M.Y.M.; Hassan, H.; Sekaran, S.D. Detection of quorum sensing signal molecules and identification of an autoinducer synthase gene among biofilm forming clinical isolates of *Acinetobacter* spp. PLoS One 2012, 7, e36696.

23. O'Toole, G.A.; Kolter, R. Initiation of biofilm formation in *Pseudomonas fluorescens* WCS365 proceeds via multiple, convergent signalling pathways: A genetic analysis. Mol. Microbiol. 1998, 8, 449–461.

24. Biswa, P.; Doble, M. Production of acylated homoserine lactone by Gram-positive bacteria isolated from marine water. FEMS Microbiol. Lett. 2013, 343, 34–41.

25. Ponnusamy, K.; Paul, D.; Kweon, J. Inhibition of quorum sensing mechanism and *Aeromonas hydrophila* biofilm formation by vanillin. Env. Eng. Sci. 2009, 26, 1359–1363.

26. Zhang, Z.; Schwartz, S.; Wagner, L.; Miller, W. A greedy algorithm for aligning DNA sequences. J. Comput. Biol. 2000, 7, 203–214.

27. Thompson, J.D.; Higgins, D.G.; Gibson, T.J. CLUSTAL W-improving the sensitivity of progressive multiple sequence alignment through sequence weighting, position specific gap penalties and weight matrix choice. Nucleic Acids Res. 1994, 22, 4673–4680.

28. Tamura, K.; Dudley, J.; Nei, M.; Kumar, S. MEGA4: Molecular evolutionary genetics analysis (MEGA) software version 4.0. Mol. Biol. Evol. 2007, 24, 1596–1599.

29. Oh, H.S.; Yeon, K.M.; Yang, C.S.; Kim, S.R.; Lee, C.H.; Park, S.Y.; Han, J.Y.; Lee, J.K. Control of membrane biofouling in MBR for wastewater treatment by quorum quenching bacteria encapsulated in microporous membrane. Environ. Sci. Technol. 2012, 46, 4877–4884.

30. Kumari, A.; Pasini, P.; Daunert, S. Detection of bacterial quorum sensing N-acyl homoserine lactones in clinical samples. Anal. Bioanal. Chem. 2008, 391, 1619–1627.

There are several supplemental files that are not available in this version of the article. To view this additional information, please use the citation on the first page of this chapter.

CHAPTER 13

Scenario Analysis of Nutrient Removal from Municipal Wastewater by Microalgal Biofilms

NADINE C. BOELEE, HARDY TEMMINK, MARCEL JANSSEN, CEES J. N. BUISMAN AND RENÉ H. WIJFFELS

13.1 INTRODUCTION

The conventional treatment of municipal wastewater consists of activated sludge processes with a combination of nitrification and denitrification and biological or chemical phosphorus removal. However, other treatment systems are also used, including systems based on microalgae, eukaryotic microorganisms and prokaryotic cyanobacteria that carry out oxygenic photosynthesis [1]. Microalgae have a high affinity for nitrogen (N) and phosphorus (P), illustrated by the low values reported for half-saturation constants, ranging from 0.56 to 3094 µg N/L, and from 0.001 to 81.9 µg P/L [2–5]. Microalgae can either grow in suspension (phytoplankton) or on substrata (benthic) in biofilms [6]. Microalgal biofilms are attached microbial consortia of phototrophs and chemotrophs entrapped in an exopolymeric matrix, and are omnipresent in aquatic environments [7,8]. Although not given a lot of attention, microalgal biofilms systems could form interesting wastewater treatment systems. A microalgal biofilm system can operate at short hydraulic retention times due to the ability of the biofilm to retain the biomass. It is also expected that, in contrast to suspended microalgal systems, little or no separation of microalgae and water is required before discharging the effluent [9,10]. Furthermore, no mixing is needed in the system, resulting in a lower energy requirement than for suspended systems.

Algal biofilms systems can be composed of large biofilm panels over which the wastewater flows. These panels can either be placed horizontally, at an angle like the algal turf scrubber [11], or vertically in rows like the twin layer system [12]. Such microalgal biofilms may be used at different stages of the wastewater treatment. A first scenario is using microalgal biofilms as a post-treatment system. In light of the EU Water Framework Directive objective to obtain good chemical and ecological status for all surface waters by 2015, this can be an interesting concept. The high affinity of microalgae for N and P and the lack of requirement of an organic carbon source are advantages over currently available post-treatment systems. A microalgal biofilm can also be used to remove N and P after a highly loaded activated sludge system. The microalgal biofilm then serves as an alternative for nitrification and denitrification and chemical or biological P removal. This scenario holds the advantages of a higher net heterotrophic biomass yield, and a lower energy input for aeration compared to a conventional wastewater treatment system.

A third option is applying an algal-bacterial biofilm to treat the wastewater directly. This scenario makes use of a symbiotic relationship that may develop when using microalgae and heterotrophs together. During this symbiosis the microalgae produce oxygen (O_2) that is needed by aerobic heterotrophs, and the carbon dioxide (C_{O2}) that is released by these heterotrophs is in turn used by the microalgae. In this manner no external O_2 supply is needed, which saves the energy otherwise required for aeration of the activated sludge system.

Previous studies have shown that microalgal biofilms systems can achieve good removal of N and P from wastewater. Removal capacities over 90% were measured for ammonium (NH_4^+), nitrate (NO_3^-) and over 80% for phosphate (PO_4^{3-}) [12–14], and up to 75% of the Chemical Oxygen Demand (COD) was removed from diluted swine manure in an algal-bacterial biofilm [14]. However, the feasibility of the application of micro-algal biofilms in wastewater treatment will be determined by more factors than the removal capacity. These factors include the achieved effluent concentrations, biomass production and the area requirement. Especially the latter is a point of concern, as algal systems are known for their relatively large area requirement. Unfortunately, little is known about these aspects of microalgal biofilms and how the three different scenarios mentioned above compare.

This study aims to get insight in the feasibility of using microalgal biofilms for municipal wastewater treatment. A scenario analysis will be performed for the three different concepts of using microalgal biofilms in municipal wastewater treatment. This analysis compares the area requirement, achieved effluent concentrations and biomass production under the conditions of municipal wastewater treatment in the Netherlands. In addition, this study seeks to determine what knowledge is still lacking in order to be able to make a final conclusion on the feasibility of microalgal biofilms in wastewater treatment.

13.2 MATERIAL AND METHODS

13.2.1 SCENARIOS

Three different scenarios were defined in which microalgae are integrated in a municipal wastewater treatment plant (WWTP), as shown in Figure 1.

In Scenario 1 the microalgal biofilm system is used as a post-treatment system for effluent from an activated sludge process. In Scenario 2 the first stage of wastewater treatment removes the bulk of the COD. Nitrification is prevented in this stage by operating at a short sludge retention time (SRT; 2.5 days). The second stage consists of a microalgal biofilm system removing N and P. In contrast to Scenario 1, N is mainly present as NH_4^+ rather than NO_3^-. In Scenario 3 the microalgae are used in a symbiotic process of algae and bacteria. N is removed by combined nitrification and denitrification and via assimilation by microalgae, COD is removed by heterotrophs, and P is mainly assimilated by microalgae.

In all scenarios the target effluent values were 2.2 mg/L total N and 0.15 mg/L total P. These values are the maximum tolerable risk (MTR) guidelines which are used by the Dutch water boards, as the classification of the good chemical and ecological status of surface water of the Water Framework Directive is not yet known.

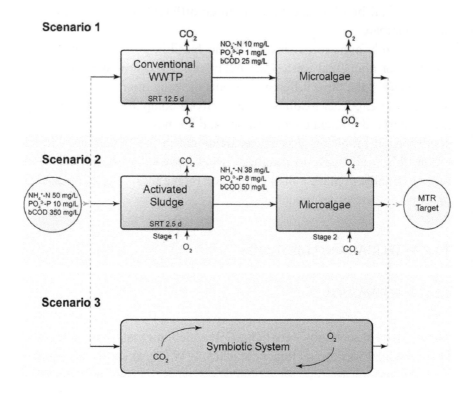

Figure 1: Schematic overview of the three different scenarios of using microalgal biofilms in municipal wastewater treatment. The incoming wastewater and target effluent MTR values of 2.2 mg/L N and 0.15 mg/L P are equal for all scenarios. The sludge retention time (SRT) is shown for the activated sludge compartments of Scenarios 1 and 2.

13.2.2 CALCULATIONS AND PARAMETERS

13.2.2.1. MICROALGAE

The WWTP in this study was located in the Netherlands and receives wastewater from 100,000 inhabitants producing 130 L per person equivalent (PE) per day. In the Netherlands, microalgal systems offer the highest potential at tourist locations during the period late spring-early autumn, when an increased wastewater production is accompanied by the highest irradiation of the year. Therefore, only the period late spring-early autumn was considered in this analysis, corresponding with the tourist season. The microalgal biofilm system therefore receives irradiation summed over the months May until October. The microalgae utilize 43% of this irradiation, equivalent to the photosynthetic active radiation (PAR, 400–700 nm). The variation of irradiance during the day and over these months has not been taken into account, but will be discussed subsequently. The summed irradiation was equal to 4773 mol photons/m^2 (see Appendix A).

An important parameter of the microalgal system is the quantum requirement of the photosynthetic process. This is the efficiency with which the microalgae take up light energy and convert it to chemical energy, i.e., new biomass, while releasing O_2. Under low light intensities, approximately 10 PAR photons are required for the liberation of one molecule of O_2 [15,16]. At higher light intensities, photosaturation takes place and part of the absorbed light is lost in the form of heat. Considering this photosaturation effect, a vertical positioning of the microalgal biofilm system is proposed and a minimal quantum requirement of 20 PAR photons per O_2 is envisioned [17]. This corresponds to a maximum oxygen quantum yield (Q_{YO2}) of 0.05 O_2/photon. With Q_{YO2} the amount of O_2 produced per photons received per m^2 of ground area is calculated:

$$R_{o,A,algae} = QY_{o_2} \cdot PFD \, [mol/m^2/d]$$

with $R_{o,A,algae}$ the areal oxygen production by microalgae [mol/m^2/d]; PFD the photon flux density [mol photons/m^2/d].

The following stoichiometrical reactions are used for microalgae, with either nitrate (assumed in Scenario 1) or ammonium (assumed in Scenarios 2 and 3) as nitrogen source.

$$CO_2+0.94H_2O+0.12NO_3^-+0.01H_2PO_4^- \rightarrow CH_{1.78}O_{0.36}N_{0.12}P_{0.01}+1.42O_2+0.13OH^-$$

$$CO_2+0.70H_2O+0.12NH_4^++0.01H_2PO_4^- \rightarrow CH_{1.78}O_{0.36}N_{0.12}P_{0.01}+1.18O_2+0.11H^+$$

The areal amount of biomass produced ($P_{x,A,algae}$) is calculated with Equation 2 or 3; 1.42 mol O_2 coincides with 1 mol of biomass (on C-basis) in case of NO_3^- uptake or 1.18 mol O_2 with 1 mol of biomass in case of NH_4^+ uptake. This biomass is assumed to be present in a biofilm kept at optimal thickness through regular harvesting of the biofilm. Keeping the biofilm at this thickness will reduce respiration losses and ensure an optimal nutrient uptake capacity.

It is expected that the microalgal biofilm will be composed of a mixed culture of microalgae, due to varying conditions with respect to temperature, and N and P concentrations. We assumed an average microalgal biomass composition for this mixed culture, of 7.8% N and 1.4% P (w/w, based on algal biomass of $CH_{1.78}O_{0.36}N_{0.12}P_{0.01}$) [18–20]. With the amount of biomass produced known and the fraction of N and P present in the biomass, the uptake of N and P from the wastewater is calculated. The following calculation shows the uptake of N, but the calculation of the P uptake is equivalent:

$$R_{N,A,algae}=R_{x,A,algae} \cdot f_{N,algae} [g/m^2/d]$$

with $R_{N,A,algae}$ the areal N uptake rate by microalgae [g/m²/d]; $P_{x,A,algae}$ the areal microalgal biomass production rate [g/m²/d] and $f_{N,algae}$ the fraction of N in the microalgal biomass [g/g].

Using the desired amount of N or P removed, the area was calculated as:

$$A=\frac{Q \cdot (N_{in}-N_{out})}{R_{N,A,algae}} [m^2]$$

with A the area [m²] and Q the flowrate [m³/d].

13.2.2.2 HETEROTROPHS

It was assumed that biomass production in the activated sludge process is only accounted for by the heterotrophic biomass converting COD. The following formula is used [21]:

$$P_{x,sludge} = Q \cdot Y_{sludge} \cdot \frac{(COD_{b,in} - COD_{b,out})}{1 + k_d \cdot SRT} \cdot (1 + f_d \cdot k_d \cdot SRT) [gVSS/d]$$

with $P_{x,sludge}$ the sludge production [g VSS/d]; Y_{sludge} the biomass yield [g VSS/g bCOD], SRT the sludge retention time [d]; f_d the fraction remaining as cell debris [g VSS/g VSS] and k_d the decay coefficient [d⁻¹].

The sludge in the WWTP was assumed to have an average composition of 12% N and 2% P (w/w based on sludge biomass of C1H1.4O0.4N0.2) [21]. With the amount of biomass produced known and the fraction of N and P present in the biomass, the uptake of N ($R_{N,sludge}$) and P from the wastewater is calculated. Calculation shown here for N:

$$R_{N,sludge} = R_{x,sludge} \cdot f_{N,sludge} [g/d]$$

Additional calculations of O_2 and CO_2 production and consumption, and a list of all parameters can be found in Appendix A.

13.3 RESULTS

Table 1 shows the area requirement of the microalgal biofilm and the corresponding effluent concentrations for the different scenarios. The area requirements are based on the calculated uptake capacities of 1.85 g N/m²/d and 0.34 g P/m²/d in Scenario 1 and 2.2 g N/m²/d and 0.41 g P/m²/d in Scenarios 2 and 3. The area requirement of the post-treatment system of Scenario 1 is the smallest with 0.32 m²/PE, followed by the symbiotic system of Scenario 3 requiring 0.76 m²/PE. The large area requirement of 2.1 m²/PE of Scenario 2 in comparison to Scenarios 1 and 3 was due to the larger amount of N that needed to be assimilated by the microalgae in this scenario.

Table 1. The required ground area and effluent concentrations of N and P of a microalgal biofilm system treating wastewater from 100,000 inhabitants in the Netherlands during May to October for the three different scenarios.

Scenario	Area requirement (m^2/PE	Effluent total N (mg/L)	Effluent total P (mg/L)
Scenario 1	0.32	5.39	0.15
Scenario 2	2.10	2.20	1.40
Scenario 3	0.76	2.20	6.07

The limiting nutrient was found to be P for Scenario 1 and N for Scenarios 2 and 3. The calculations were therefore performed for P reaching the desired MTR value in Scenario 1 and for N reaching the desired value for Scenarios 2 and 3. Consequently, in the P limiting Scenario 1, the N concentration remained above target with 5.39 mg N/L, while in the N limiting Scenarios 2 and 3 the P concentration remained above target. The latter two also did not comply with current EU effluent discharge requirements of 1 mg P/L. Consequently, the desired effluent values for both N and P could not be reached simultaneously.

In Scenario 3, a symbiotic relationship between microalgae and heterotrophs was assumed to develop. The O_2 production by the microalgae and the O_2 consumption by the heterotrophs were balanced, by adjusting the fraction of N that was removed by combined nitrification and denitrification and the fraction that was assimilated by microalgae. Figure 2 shows that the O_2 in the system was balanced when 70% of the NH_4^+ was converted by the heterotrophs (nitrification-denitrification) and the remaining 30% by the microalgae. With this balance, the microalgae supply all O_2 for the heterotrophs, and aeration is theoretically not needed. However, it can also be seen from Figure 2 that the CO_2 production and consumption could not be balanced at the same time. Approximately 40% additional CO_2 needs to be supplied or fixed by the microalgae from the air.

Figure 3 shows the amount of activated sludge and microalgal biomass produced in the three different scenarios. The microalgal biomass per PE was based on the calculated microalgal biomass production of 24 $g/m^2/d$ in Scenario 1 and 28 $g/m^2/d$ in Scenarios 2 and 3. In Scenario 1 similar amounts of activated sludge and microalgal biomass were produced. Scenario 2 had the largest microalgal biomass production of 59 g/PE/d, because larger amounts of nutrients needed to be assimilated in this scenario.

13.4 DISCUSSION

13.4.1 EFFLUENT CONCENTRATIONS

The results of this scenario analysis showed that it was not possible to simultaneously remove the N and P in the wastewater to the target values of 2.2 mg N/L and 0.15 mg P/L. In the post-treatment scenario P, while in Scenarios 2 and 3 N was limiting the microalgae growth. Indeed, given the ratio of N:P in wastewater, N will always be the limiting nutrient, if the molar ratio of C:N:P of 100:12:1 represents the real average elemental composition of microalgae grown in such systems. However, elemental composition in microalgae is known to be highly variable. Compositions with C:P ratios between 34:1 and 418:1, and N:P ratios between 3.5:1 and 38:1 are reported in literature for different species of microalgae [20,22]. Also the growth conditions, with respect to nutrient and/or light limitation, influence the elemental composition. Molar N:P ratios as low as 3:1 have been reported under N limiting conditions, while under conditions of P limitation a N:P ratio of 100:1 is possible [23,24]. At low growth rates, the N:P ratio of the biomass composition can sometimes match the supply ratio. Especially luxury uptake of P with storage as polyphosphate, is known to take place in microalgae [25,26]. This luxury uptake might make it possible to not only achieve MTR quality with respect to N, but also with respect to P.

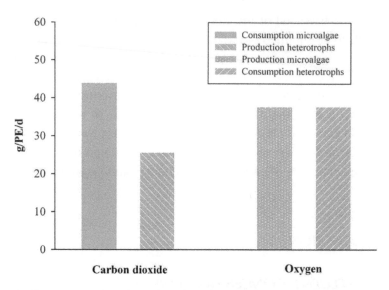

Figure 2: The consumption of CO_2 by microalgae alongside the production of CO_2 by heterotrophs, and the production of O_2 by microalgae alongside the consumption of O_2 by heterotrophs in the symbiotic system of Scenario 3. The amounts are expressed in gram per person equivalent (PE) per day.

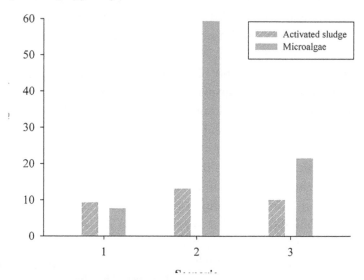

Figure 3: The microalgal biomass and activated sludge in the three scenarios for a WWTP of 100,000 inhabitants in the Netherlands from May to October. The amount of activated sludge is expressed in grams volatile suspended solids (VSS) and the amount of microalgae in grams dry weight, both per person equivalent (PE) per day.

In the scenario analysis only N and P removal by microalgal assimilation was taken into account. Additional removal of P by precipitation with cations like Ca^{2+} and Mg^{2+} is also possible. This precipitation occurs at higher pH levels caused by microalgae changing the $CO_2/HCO_3^-/CO_3^{2-}$ equilibrium when more CO_2 is taken up than can be supplied via absorption from the atmosphere [9,26]. With such a pH rise expected within the biofilm, it is likely that in practice a larger P removal occurs than was calculated.

13.4.2 AREA REQUIREMENT

This study found area requirements for the three scenarios between 0.32 and 2.1 m^2/PE. A conventional WWTP is estimated to have an area requirement around 0.2–0.4 m^2/PE. Hence, with 0.32 m^2/PE, the microalgal post-treatment requires a similar area to that of the activated sludge plant. The two-stage system of Scenario 2 requires the largest area of 2.1 m^2/PE. However, the first activated sludge stage will be considerably smaller than in a conventional WWTP, because nitrification is absent. In addition, more sludge is produced, and this sludge will have a higher energy value. This implies that more methane can be produced when digesting the sludge anaerobically.

The symbiotic system of Scenario 3 has an area requirement of 0.76 m^2/PE, which is smaller than the area of the system of Scenario 2, and similar to the conventional WWTP combined with microalgal post-treatment in Scenario 1. Moreover, this system has the advantage that the O_2 production by microalgae and O_2 consumption by heterotrophs can be balanced. This balance implies no need for an external oxygen supply, giving energy and cost savings. Although very attractive, the technology required to support a symbiotic biofilm system still needs to be developed.

The area requirements are based on the calculated uptake capacities of 1.85 g $N/m^2/d$ and 0.34 g $P/m^2/d$ in the post-treatment scenario and 2.2 g $N/m^2/d$ and 0.41 g $P/m^2/d$ in Scenarios 2 and 3. These calculated uptake capacities are higher than the 0.1–0.6 g $N/m^2/d$ and 0.006–0.09 g $P/m^2/d$ measured in lab-scale biofilm systems [13,14], but lower than the 0.7–2.1 g $P/m^2/d$ measured in other pilot scale microalgal biofilms systems

[11,27]. This indicates that the calculated uptake capacities in this study can be considered a good average estimation.

The area requirement calculated for the different microalgal systems depends on the elemental composition of microalgal biomass, as well as on the irradiance and the photosynthetic efficiency. As mentioned above, the elemental composition can vary in microalgae. Clearly a higher N and P content in the microalgae will result in a lower area requirement, while a lower content will increase the required area. This again illustrates the importance of knowing the real elemental composition of the microalgae when growing on wastewater.

The effect of irradiance on the area requirement can only be changed by either moving the system to another location, or by applying artificial illumination. However, the addition of artificial light yields no substantial area reduction. If all produced biomass, both activated sludge as well as microalgal biomass, would be converted into biogas to produce electricity for artificial light, the area reduction for the three scenarios is at most 0.8% (see Appendix B). This extremely low reduction is related to the energy losses in the process of methanogenesis, in converting biogas into electricity, electricity into light and light into new biomass via photosynthesis. Clearly these large scale microalgae based processes can only be fueled by sunlight.

In this analysis the oxygen quantum efficiency was assumed to be 0.05 mol O_2/mol photons. This value is not reached in horizontal microalgal systems, but has been reached in vertical panel photobioreactors [17]. This efficiency has been determined with light as the limiting substrate. However, it is likely that CO_2 limitation will occur when the biofilm is only exposed to ambient air. Modeling of microalgal biofilms has shown that CO_2 limitation can easily occur, the level of limitation depending on the bulk pH and alkalinity [28]. Therefore, the assumed efficiency can only be reached in a vertical biofilm system, if CO_2 limitation can be prevented. Using heterotrophic microorganisms to directly supply the CO_2 to the microalgae as in Scenario 3, may therefore be a very attractive way to accomplish this.

13.4.3 SEASONAL VARIATION IN TEMPERATURE AND LIGHT INTENSITY

Both microalgal growth and uptake of N and P decrease at lower temperatures and light intensity [29]. Therefore, the capacity of the microalgal system will change throughout the day and the seasons. Low uptake of nutrients by microalgae during winter may be one of the main limitations of the application of microalgal wastewater treatment systems in a country like the Netherlands. In this scenario analysis, the system was assumed to be running only in the tourist season, from May until October. The application of the microalgal system in places where a much higher capacity is needed during summer is the most interesting application of this technology in The Netherlands. The microalgal system will provide additional capacity during summer, whereas during winter the existing WWTP capacity will be sufficient. Such a microalgal system may be applied on the islands in the Wadden Sea of the Netherlands, being a tourist location during summer. On the island Ameland for example, wastewater production during summer can be more than three times the amount during the rest of the year. With these conditions a microalgal biofilm system is an interesting option to treat the additional wastewater in summer, instead of increasing the size of the wastewater treatment plant to be able to treat the summer wastewater load.

13.4.4 DAILY VARIATION IN LIGHT INTENSITY

The microalgae production was based on the total irradiation received in five months, and therefore the diurnal light cycle was not taken into account. In general, uptake of N and P by microalgae changes throughout the day and is faster during daytime than in the dark [24,25]. N-limited microalgae, on the other hand, are known to take up either NO_3 or NH_4^+ in the dark. Uptake of NH_4^+ can be more than 50% of the daylight value in

N-limited microalgae [30]. Consequently, when N is the limiting nutrient for microalgae in the treatment of municipal wastewater, considerable uptake of N during darkness might be expected. In addition, the wastewater loading rate of N and P is expected to be lower during the night, possibly compensating for the reduced nutrient uptake.

13.4.5 APPLICATION OF MICROALGAL BIOMASS

When using microalgal biofilm systems in wastewater treatment, substantial amounts of microalgal biomass are produced. In Scenario 1 the microalgal biomass production was 24 g/m²/d based on the consumption of NO_3^- and in Scenarios 2 and 3 the production was 28 g/m²/d based on th consumption of NH_4^+. These biomass productions are slightly higher than biomass production of 11–18 g/m²/d measured in pilot photobioreactors [31,32] and in the range of biomass production of 24–31 g/m²/d measured in other biofilm pilot systems [11,27]. With this biomass production, a WWTP for 100,000 inhabitants with the microalgal post-treatment of Scenario 1 produces 200 ton (dry weight) microalgal biomass during the five summer months of operation. In Scenarios 2 and 3 this amount of biomass is even larger. With such a substantial amount of biomass, it is important to find an efficient way to harvest the biomass, and a proper destination.

To ensure an actively growing biofilm, and to prevent washout of valuable biomass with the effluent, the biofilm will need to be harvested regularly. This regular harvesting will reduce respiratory biomass losses or even cell death, which otherwise would lead to release of N and P from the biofilm. Two ways of harvesting can be distinguished, passive and active. Passive harvesting entails collecting the microalgal biomass that naturally detaches from the top of the biofilm when it ages. This might involve the addition of a settler tank, resulting in extra area requirement. Active harvesting techniques currently applied to remove biofilms in other systems include pH shock [33], backwashing and scraping. Active harvesting appears more attractive as it gives the possibility to harvest very regularly, hereby reducing the respiratory losses as much as possible.

Nitrogen was shown to be the limiting nutrient when integrating microalgal biofilms in the wastewater treatment (Scenarios 2 and 3). In this case, it might be possible to accumulate lipids in the microalgal biomass, as microalgae start to accumulate these under conditions of N limitation [34,35]. To achieve this N-starvation and induce lipid accumulation the C:N ratio should be twice as high as was assumed in the scenario analysis. Although this increased ratio would result in an area requirement twice as large, the amount of produced biomass will also be doubled and thus larger amounts of lipids may be produced. In Scenario 1 these lipids would approximately amount to 79 ton during the five summer months of operation (see Appendix C). However, further research is still needed to induce the accumulation of specific desired lipids, to obtain a stable (mixed) culture of the desired species in the system, and to set up the biorefinery needed to extracts these lipids as well as valorize the remaining biomass constituent [30,31].

Depending on the microalgal biomass composition other products might also be possible. Using the biomass as fertilizer is very attractive but is only possible when no heavy metals or other recalcitrant compounds are present in the wastewater and accumulated by microalgae. Anaerobic digestion for biogas production is another possibility, although afterwards still autotrophic N and P removal or recovery will be necessary. The CO_2 that is produced during digestion might be recycled to the microalgal biofilm system as an additional CO_2 supply [36,37].

13.5 CONCLUSIONS

This study investigated the potential of a hypothetical microalgal biofilm system as a seasonal wastewater treatment system in the Netherlands. The analysis showed that the area requirement of the microalgal biofilm system was 0.32 m^2/PE for a post-treatment system, 2.10 m^2/PE for a two stage wastewater treatment system and 0.76 m^2/PE for a one-stage symbiotic system. In addition, it was found that microalgae growing on wastewater treatment plant effluent are P limited and microalgae growing on untreated or partially treated wastewater are N limited. The microalgae will produce a substantial amount of biomass. For the application of microalgal biofilms in countries such as the Netherlands, further research should look into the effect of the daily variation of both the wastewater flows and of the irradiation and temperature. In addition, the destination of the produced biomass is an important topic for future studies. Finally, real (pilot) tests should be performed to establish if indeed a photosynthetic efficiency of 0.05 mol O_2/mol photons can be reached and whether CO_2 limitation will occur.

REFERENCES

1. Brock, T.D.; Madigan, M.T.; Martinko, J.M.; Parker, J. Biology of Microorganisms, 9th ed.; Prentice-Hall: Upper Saddle River, NJ, USA, 2000.
2. Halterman, S.G.; Toetz, D.W. Kinetics of nitrate uptake by freshwater algae. Hydrobiologia 1984, 114, 209–214.
3. Collos, Y.; Vaquer, A.; Souchu, P. Acclimation of nitrate uptake by phytoplankton to high substrate levels. J. Phycol. 2005, 41, 466–478.
4. Eppley, R.W.; Rogers, J.N.; McCarthy, J.J. Half-Saturation constants for uptake of nitrate and ammonium by marine phytoplankton. Limnol. Oceanogr. 1969, 14, 912–920.
5. Hwang, S.J.; Havens, K.E.; Steinman, A.D. Phosphorus kinetics of planktonic and benthic assemblages in a shallow subtropical lake. Freshw. Biol. 1998, 40, 729–745.
6. Stevenson, R.J.; Bothwell, M.L.; Lowe, R.L. Algal Ecology: Freshwater Benthic Ecosystems; Academic Press Elsevier: San diego, CA, USA, 1996.
7. Di Pippo, F.; Ellwood, N.; Guzzon, A.; Siliato, L.; Micheletti, E.; de Philippis, R.; Albertano, P. Effect of light and temperature on biomass, photosynthesis and capsular polysaccharides in cultured phototrophic biofilms. J. Appl. Phycol. 2011, 24, 1–10.
8. Sekar, R.; Nair, K.V.K.; Rao, V.N.R.; Venugopalan, V.P. Nutrient dynamics and successional changes in a lentic freshwater biofilm. Freshwater Biol. 2002, 47, 1893–1907.
9. Roeselers, G.; Loosdrecht, M.; Muyzer, G. Phototrophic biofilms and their potential applications. J. Appl. Phycol. 2008, 20, 227–235.
10. Schumacher, G.; Blume, T.; Sekoulov, I. Bacteria reduction and nutrient removal in small wastewater treatment plants by an algal biofilm. Water Sci. Technol. 2003, 47, 195–202.
11. Craggs, R.J.; Adey, W.H.; Jenson, K.R.; St. John, M.S.; Green, F.B.; Oswald, W.J. Phosphorous removal from wastewater using algal turf scrubber. Water Sci. Technol. 1996, 33, 191–198.
12. Shi, J.; Podola, B.; Melkonian, M. Removal of nitrogen and phosphorus from wastewater using microalgae immobilized on twin layers: An experimental study. J. Appl. Phycol. 2007, 19, 417–423.

13. De Godos, I.; González, C.; Becares, E.; García-Encina, P.; Muñoz, R. Simultaneous nutrients and carbon removal during pretreated swine slurry degradation in a tubular biofilm photobioreactor. Appl. Microbiol. Biotechnol. 2009, 82, 187–194

14. González, C.; Marciniak, J.; Villaverde, S.; León, C.; García, P.A.; Munoz, R. Efficient nutrient removal from swine manure in a tubular biofilm photo-bioreactor using algae-bacteria consortia. Water Sci. Technol. 2008, 58, 95–102.

15. Ley, A.C.; Mauzerall, D.C. Absolute absorption cross-sections for Photosystem II and the minimum quantum requirement for photosynthesis in *Chlorella vulgaris*. BBA—Bioenergetics 1982, 680, 95–106.

16. Bjorkman, O.; Demmig, B. Photon yield of O2 evolution and chlorophyll fluorescence characteristics at 77 K among vascular plants of diverse origins. Planta 1987, 170, 489–504.

17. Qiang, H.; Faiman, D.; Richmond, A. Optimal tilt angles of enclosed reactors for growing photoautotrophic microorganisms outdoors. J. Ferment. Bioeng. 1998, 85, 230–236.

18. Duboc, P.; Marison, I.; Stockar, U.V. Quantitative calorimetry and biochemical engineering. In Handbook of Thermal Analysis and Calorimetry; Kemp, R.B., Ed.; Elsevier: Amsterdam, The Netherlands, 1999; Volume 4.

19. Healey, F.P. Inorganic nutrient uptake and deficiency in algae. Crit. rev. microbiol. 1973, 3, 69–113.

20. Ahlgren, G.; Gustafsson, I.B.; Boberg, M. Fatty-acid content and chemical—Composition of fresh-water microalgae. J. Phycol. 1992, 28, 37–50.

21. Metcalf, I.; Eddy, H. Wastewater Engineering: Treatment and Reuse, 4th ed.; McGraw-Hill: Columbus, OH, USA, 2003.

22. Ho, T.Y.; Quigg, A.; Finkel, Z.V.; Milligan, A.J.; Wyman, K.; Falkowski, P.G.; Morel, F.M.M. The elemental composition of some marine phytoplankton. J. Phycol. 2003, 39, 1145–1159.

23. Goldman, J.C.; McCarthy, J.J.; Peavey, D.G. Growth rate influence on the chemical composition of phytoplankton in oceanic waters. Nature 1979, 279, 210–215.

24. Elrifi, I.R.; Turpin, D.H. Steady-state luxury consumption and the concept of optimum nutrient ratios: A study with phosphate and nitrate limited *Selenastrum minutum* (Chlorophyta). J. Phycol. 1985, 21, 592–602.

25. Klausmeier, C.A.; Litchman, E.; Simon, A.L. Phytoplankton growth and stoichiometry under multiple nutrient limitation. Limnol. Oceanogr. 2004, 49, 1463–1470.

26. Powell, N.; Shilton, A.N.; Pratt, S.; Chisti, Y. Factors Influencing luxury uptake of phosphorus by microalgae in waste stabilization ponds. Envir. Sci. Technol. 2008, 42, 5958–5962.

27. Christenson, L.B.; Sims, R.C. Rotating algal biofilm reactor and spool harvester for wastewater treatment with biofuels by-products. Biotechnol. Bioeng. 2012, doi: 10.1002/bit.24451.

28. Liehr, S.K.; Eheart, J.W.; Suidan, M.T. A modeling study of the effect of pH on carbon limited algal biofilms. Water Res. 1988, 22, 1033–1041.

29. Goldman, J.C.; Carpenter, E.J. A Kinetic approach to the effect of temperature on algal growth. Limnol. Oceanogr. 1974, 19, 756–766.

30. Vona, V.; Rigano, V.D.M.; Esposito, S.; Carillo, P.; Carfagna, S.; Rigano, C. Growth, photosynthesis, and respiration of *Chlorella sorokiniana* after N-starvation. Interactions between light, CO_2 and NH_4^+ supply. Physiol. Plantarum 1999, 105, 288–293.

31. Hulatt, C.J.; Thomas, D.N. Energy efficiency of an outdoor microalgal photobioreactor sited at mid-temperature latitude. Bioresource technol. 2011, 102, 6687–6695.

32. Min, M.; Wang, L.; Li, Y.; Mohr, M.J.; Hu, B.; Zhou, W.; Chen, P.; Ruan, R. Cultivating *Chlorella* sp. in a pilot-scale photobioreactor using centrate wastewater for microalga biomass production and wastewater nutrient removal. Appl. Biochem. Biotech. 2011, 165, 123–137.

33. 33. Knuckey, R.M.; Brown, M.R.; Robert, R.; Frampton, D.M.F. Production of microalgal concentrates by flocculation and their assessment as aquaculture feeds. Aquacult. Eng. 2006, 35, 300–313.

34. Converti, A.; Casazza, A.A.; Ortiz, E.Y.; Perego, P.; del Borghi, M. Effect of temperature and nitrogen concentration on the growth and lipid content of *Nannochloropsis oculata* and *Chlorella vulgaris* for biodiesel production. Chem. Eng. Process. 2009, 48, 1146–1151.

35. Solovchenko, A.; Khozin-Goldberg, I.; Didi-Cohen, S.; Cohen, Z.; Merzlyak, M. Effects of light intensity and nitrogen starvation on growth, total fatty acids and arachidonic acid in the green microalga *Parietochloris incisa*. J. Appl. Phycol. 2008, 20, 245–251.

36. Muñoz, R.; Guieysse, B. Algal-bacterial processes for the treatment of hazardous contaminants: A review. Water Res. 2006, 40, 2799–2815.

37. Mussgnug, J.H.; Klassen, V.; Schlüter, A.; Kruse, O. Microalgae as substrates for fermentative biogas production in a combined biorefinery concept. J. Biotechnol. 2010, 150, 51–56.

38. Wetsus Centre of Excellence for Sustainable Water Technology Home Page. Available online: http://www.wetsus.nl (accessed on 16 April 2012).

There are several supplemental files that are not available in this version of the article. To view this additional information, please use the citation on the first page of this chapter.

Author Notes

CHAPTER 1

Acknowledgments

Authors wish to thank the Natural Science Foundation of China (21377123 and 51322802), Hefei Center for Physical Science and Technology (2012FXZY005) and the Program for Changjiang Scholars and Innovative Research Team in University of Ministry of Education of China for the partial support of this study. Authors also wish to thank the Shanghai Synchrotron Radiation Facility, Shanghai, China for XAFS analysis.

Author contributions

H.W.L. and J.J.C. carried out the experiments, analyzed the data, and wrote the paper; G.P.S. designed the experiments, analyzed the data, and wrote the paper; J.H.S. designed the EPR experiments and analyzed EPR data, and S.Q.W. analyzed the XAFS data; H.Q.Y. analyzed the data and wrote the paper.

Competing financial interests

The authors declare no competing financial interests.

CHAPTER 2

Acknowledgments

This study was financially supported by Natural Science Foundation of China (51208009 and 21177005) and Natural Science Foundation of Beijing

(8132008). We also acknowledge the support from Specialized Research Fund for the Doctoral Program of Higher Education (20121103120010) and National High Technology Research and Development Program (863 Program) of China (2011AA060903-02). J. Guo acknowledges the support of the Australian Research Council Discovery Early Career Researcher Award (DE13010140). B.-J. Ni acknowledges the support of the Australian Research Council Discovery Project (DP130103147).

Author contributions
J.G. and Y.P. conceived and designed the experiments; A.Z. and J.G. performed the modelling and analyzed the data; S.W., B.-J.N., Q.Y. contributed materials/analysis tools; A.Z., J.G. and B.-J.N. wrote the paper. All authors reviewed the manuscript.

Competing financial interests
The authors declare no competing financial interests.

CHAPTER 3

Funding
The project was funded by the Swedish Research Council Formas (http://formas.se/), grant number 2012–1433. The funder had no role in study design, data collection and analysis, decision to publish, or preparation of the manuscript.

Competing interests
The authors have declared that no competing interests exist.

CHAPTER 4

Acknowledgments
This study was financially supported by the Hong Kong General Research Fund (HKU7202/09E) and Bing Li thanks The University of Hong Kong for the postgraduate studentship.

CHAPTER 5

Funding
JJ is thankful to the Council of Scientific and Industrial Research (http://www.csirhrdg.res.in/), Government of India, for her fellowship. RK is thankful to the Department of Biotechnology (http://www. dbtindia.nic. in/index.asp), Government of India, for his fellowship. The authors are thankful to Council of Scientific and Industrial Research (Sustainable Environmental Technologies for Chemical and Allied Industries: CSC-0113) for funding that aided the study design, data collection and analysis.

Competing interests
The authors have declared that no competing interests exist.

CHAPTER 6

Competing interests
The authors declare that there are no competing interests.

Authors' contributions

BH drafted the manuscript, designed and carried out the biodegradation experiments. HL reviewed the manuscript. HH and EM conceived of the study, participated in its coordination and helped to review the manuscript. All authors read and approved the final manuscript.

Acknowledgement

Financial support by the Bavarian State Ministry of the Environment and Public Health (StMUG) is gratefully acknowledged.

CHAPTER 7

Funding information

This work was supported by the Spanish projects Consolider TRAGUA (CSD2006-00044) and CTQ2009- 14390-C02-02.

Acknowledgements

We thank the Bioinformatics Platform UAB (BioinfoUAB), Ramiro Logares and Guillem Salazar for help and support with sequence analyses.

Conflict of interest

None declared.

CHAPTER 8

Acknowledgments

This research was financially supported by the project of National Natural Science Foundation of China (NSFC) (Nos. 51278175 and 51378188), International Science & Technology Cooperation Program of China (No.

2012DFB30030-03), Hunan Provincial Innovation Foundation for Post-graduate (CX2014B137), and National Science Foundation of Jiangsu Province (BK2012253).

Author contributions
J.W.Z. carried out the experiments and drafted the paper, D.B.W. and X.M.L. designed the experimental plan and revised the paper, Q.Y., H.B.C., Y.Z., H.X.A. and G.M.Z. analyzed the data. All authors contributed to the scientific discussion.

Competing financial interests
The authors declare no competing financial interests.

CHAPTER 9

Acknowledgments
Financial support was provided by the Polish Ministry of Science and Higher Education Grant No.: N305 320 636 in the years 2009–2011 and N305 327 439 in years 2010–2011 and by the European Union within the European Social Fund in project "InnoDoktorant—Scholarships for PhD students, 2nd edition". Help of Ms. Iwona Henke in performing strain isolation is greatly acknowledged. We would like to thank Daniel Szopinski for the artwork. In addition, we would like to thank Aleksandra Markiewicz for her assistance with bioinformatics.

CHAPTER 10

Acknowledgments
This study was supported by the Australian Research Council (ARC) through Project DP130103147. Yiwen Liu gratefully received the En-

deavour International Postgraduate Research Scholarship (IPRS) and The University of Queensland Centennial Scholarship (UQCent). Bing-Jie Ni acknowledges the supports of ARC Discovery Early Career Researcher Award (DE130100451) and ARC Linkage Project (LP110201095). Yaobin Zhang acknowledges the supports of National Natural Scientific Foundation of China (51378087 and 21177015).

Author contributions
Y.L., Q.W., Y.Z. and B.-J.N. wrote the manuscript; Y.L., Y.Z. and B.-J.N. developed the methodology; Y.L. and Q.W. performed data analysis and prepared all figures; All authors reviewed the manuscript.

Additional information
Competing financial interests: The authors declare no competing financial interests.

CHAPTER 11

Competing interests
The authors declare that they have no competing interests.

Authors' contributions
MTW performed the sample preparation, sequence analyses, data interpretation and compiled the manuscript. DZ carried out the anaerobic digestion and physical/chemical characterization. JL compiled the computational analyses tools. RKHH performed the next-generation sequencing. HMT, MSB and TJP contributed to data interpretation and manuscript's final revision. FCL and YC conceived and coordinated the study. All authors read and approved the final manuscript.

Acknowledgements
This study was partially supported by Initiative on Clean Energy and Environment, The University of Hong Kong (HKU-ICEE) and a Bioinformat-

ics Center fund from Nanjing Agricultural University. We are grateful to P. Benzie, C. Bush and I. Tasovski for editing the manuscript

Competing interests
The authors declare that they have no competing interests.

CHAPTER 12

Acknowledgments
This research was supported by the National Research Foundation of Korea (NRF) grant funded by the Korean Government (MSIP) to Konkuk University, Seoul, Korea (No. 2012034725). A project with Seoul National University (Seoul, Korea) supported by Korea Ministry of Environment is also acknowledged (2012001440001). We gratefully acknowledge Chung-Hak Lee, School of Chemical and Biological Engineering, Seoul National University (Seoul, Korea) for providing QS-reporter strain A. tumefaciens A136 and Hee Deung Park, School of Civil, Environmental and Architectural Engineering, Korea University (Seoul, Korea) for providing QS-reporter strain *C. violaceum* CV026.

Conflicts of interest
The authors declare no conflict of interest.

CHAPTER 13

Acknowledgements
This work was performed in the TTIW-cooperation framework of Wetsus, Centre of Excellence for Sustainable Water Technology [38]. Wetsus is funded by the Dutch Ministry of Economic Affairs, the European Union Regional Development Fund, the Province of Fryslân, the City of

Leeuwarden and the EZ/Kompas program of the "Samenwerkingsverband Noord-Nederland". The authors like to thank the participants of the research theme "Advanced waste water treatment" and the steering committee of STOWA for the discussions and their financial support.

Index

A

A. tumefaciens 271–274, 278, 280, 283–285, 288, 319
acetate 16, 107, 112, 130–131, 152–153, 188, 191, 196, 238, 244–246, 248–249, 252, 263, 278, 281, 287
acetogenesis 219
acidification 223
Actinobacteria 123, 125, 141–142, 146, 165, 242, 247, 270
Actinomycetales 125
activated sludge models 23, 36, 45, 47, 69
    ASMs 23–25, 38, 45
adsorption 4, 16, 20, 22, 50–51, 59–60, 71–72, 76–87, 89–101, 208
    adsorption affinity 93, 98
    adsorption-biooxidation 50
Aeromonas 270, 276–278, 290, 292
AHLs 270–290
    sludge AHLs 270–271, 278, 282, 288–289
Alphaproteobacteria 125, 135, 164
ammonia oxidization 167
ammonium 28, 34, 36, 43, 91, 160, 171, 216, 284, 294, 298, 308
amoA viii, 159, 161, 163, 166–169, 171
anaerobic digester 50–51, 56–57, 225, 228, 231, 240, 242, 254–255, 258, 262
antibiotics 75, 78, 98–100, 137–138, 149, 155–156, 158, 284–285, 291

antibiotic resistance 99, 138
Archaea 168, 170–172, 238, 241, 244–246, 249, 262
ASM3 23, 25, 30–31, 34, 37, 42, 44, 72
automatic calibration 38
autotrophic organisms 31
azide 54, 60–61, 65–66, 69, 72, 80
Azoarcus 125
Azoarocous 126
Azonexus 125–126

B

bacteria ix, 21, 43, 75, 98–99, 106–107, 111, 122–123, 130, 133, 135, 138, 147, 152, 155–157, 162, 164, 167–172, 182, 195, 197, 202–203, 206, 209–211, 215–216, 238, 241–242, 244–245, 260, 262–263, 269–273, 275–277, 279–281, 283–287, 289–292, 295, 309
Bacteroidetes 161, 163, 165, 169, 242, 244, 247, 270
Betaproteobacteria 125, 161, 167–168, 270
binding parameters 10
binding sites 5–6, 21
biofilm ix, 267, 269–271, 275, 277, 286, 289–292, 294–295, 297–298, 300, 303–310
biogas 50–51, 69, 225, 228, 231, 238–240, 242–244, 253–254, 261–263, 304, 307, 310

biological nutrient removal 108, 133,
  177, 197, 244, 262
    BNR 177–179, 190–192
biological phosphorus removal viii,
  xix, 132, 134–135, 175–176, 179,
  195–197
biomass 4, 31, 37, 112, 115, 117,
  131, 142–143, 150, 152–154, 201,
  203, 214, 237, 244, 250, 261, 264,
  269, 294–295, 297–299, 301–304,
  306–308, 310
biosorption 15–16, 20, 60, 71–72
bond distances 4, 11, 13, 16
Brachymonas 125
Brevundimonas 140–143, 146, 149
buffering 5
Burkholderiales 125

C

C. violaceum 271–274, 278, 280–281,
  283–285, 288, 291, 319
calibration vii, 18, 23–26, 28–32, 34,
  36–39, 43–47, 55, 83, 153
calibration protocols 24, 45
CER 16
chemical oxygen demand 53, 72, 161,
  191, 230, 295
    COD vii, 28, 31, 34, 38–39, 51,
      53, 68, 70, 72, 78, 105, 107–109,
      111–119, 121, 123, 125, 127–129,
      131–135, 191, 193–194, 255,
      295–296, 299
Chloroflexi 163–165, 169, 242, 244,
  247, 262
Citrobacter 270, 276–278
Comamonadaceae 125
complexing mechanisms 9
conductivity 52, 54–55, 78, 231
consortiums viii, 107, 112, 116, 128,
  130, 151, 201, 203, 205–207, 209,
  211–215, 238, 247, 253, 262

contact-stabilization 49–51
continuous stirred-tank reactor 25
    CSTR 25, 28, 30–37, 39, 43–44
copper vii, 3, 5, 7–9, 11, 13, 15,
  17–21, 100, 133, 197
Corynebacteriaceae 125
Corynebacterium 125, 128
costs xxiii, 71, 139, 177, 220, 225,
  228–230, 233, 235, 304
crystallinity 120
cyanobacteria 293

D

deflocculation 54, 60, 70
Deinoccocus-Thermus 161
denitrification xix, 31, 34, 42, 44, 51,
  115–116, 125, 128, 132–133, 135,
  167, 170, 176, 182, 185, 195–196,
  293–294, 296, 301
density functional theory 5, 21
    DFT 5, 8, 13, 15–17, 19–20
dissolved organic carbon 52–53, 55,
  72, 78
    TOCd 52–53, 55–66, 68
dissolved oxygen 28, 47, 68, 181
    DO xxiii–xxiv, 28, 30–31, 36, 43,
      132, 181–182, 185–186, 211, 271,
      280, 284
Dyella 270

E

ecosystems x–xi, xix–xxvii, 39, 121,
  135, 137, 308
effluent 16, 28, 49–54, 56–58, 62, 76,
  107, 138, 178–181, 184–185, 194,
  228, 294–296, 300–301, 306, 308
electron paramagnetic resonance 4, 21
    EPR 4, 7–9, 11, 13, 15, 17, 19, 21,
      313
energy consumption 23, 44, 71

enhanced biological phosphate removal 105, 195
   EBPR 15, 105–106, 119, 123, 134, 195–196
Enterobacter 270, 276–278, 290
enzymes 98, 188, 215, 223, 238–239, 246, 248, 250–251, 261, 263–264
error propagation 24
E. coli 208
Eukaryota 241, 245
eutrophication 175
EXAFS 11–12, 18, 20–22
extracellular polymeric substances 3, 20–22, 99–100
   EPS 3–4, 9–11, 13–14, 16–17, 19, 22, 99–100, 250

F

fermentation 64, 219, 234, 244, 250, 255, 260
Firmicutes 128, 165, 242, 244, 247
Fisher Information Matrix 41, 45
Flavobacterium 207–208
fluorescence in situ hybridization 160, 194
   FISH 132, 155, 160–161, 164, 167, 169, 179, 183, 194, 291
Fourier transformations 11
free nitrous acid viii, 175–176, 195–196, 233–234
   FNA 176–181, 183–185, 188–190, 193–195, 233
Freundlich model 96
FTIR analysis 120, 128
functional groups 3–6, 9, 11, 17–18, 20, 61, 64, 91, 93, 120

G

Gammaproteobacteria 125, 128, 141, 160–161, 165, 167, 270

genetic algorithm 25, 39, 47
   GA 25, 40, 132–133
genome 47, 239–240, 248, 261, 263, 265
geochemical 3, 15
global optimization 25, 27, 38–40, 47
   GO 25, 27, 34, 36
global sensitivity measures 25, 40, 47
   DGSM 25–26, 36, 40, 42
glucose 13, 15, 19, 152, 189–190, 284
glycogen accumulating organisms 112, 132, 179
   GAO 100, 196
gram-negative 208–210, 291
gram-positive 208–210, 292

H

heavy metals xxi, 3–5, 15, 20, 99, 307
heterotrophs 31, 112, 128, 168–171, 270, 272, 294, 296, 299, 301–302, 304–305
highest occupied molecular orbital 13
   HOMO 13
hydrolysis viii, 20, 42, 76–78, 83, 97, 106, 119, 188, 219–220, 222–223, 226, 230–232, 234–235, 238–239, 248, 250–251, 283
hydrophobic 88
hydroxyl groups 13, 202

I

imidazolium viii, 201–203, 205, 207, 209, 211–213, 215–216
influent 24, 30 31, 37, 50, 52, 55–56, 68, 71, 131, 151, 161, 165, 180–181, 184, 189, 191, 234
infrared spectroscopy 4, 109
inoculum 43, 108, 202–203, 205, 211, 231, 242, 253

J

Jahn-Teller effect 8–9

K

kinetics vii, 1, 4, 21, 24, 28, 43, 47,
  51, 53–54, 60–61, 63–66, 69–70, 72,
  75, 77, 79, 81, 83–87, 89, 91, 93, 95,
  97–99, 116, 128, 130, 156, 196–197,
  231–233, 291, 308, 310
  pseudo-second-order kinetics 84,
  86, 98
Klebsiella 270, 276–278

L

Langmuir model 79–80, 96
ligands 9, 13

M

macroscopic interactions? 15
Malikia 270, 276
membrane bioreactors 45, 269,
  290–291
  MBR 44, 69, 269–272, 278, 282,
  284, 287, 290, 292
membrane
  fouling 269–271, 290
  permeabilization 208
metals xxi, 3–5, 10, 15, 20, 99, 182,
  192–193, 201, 307
methanogenesis 219, 238, 246, 248,
  252–253, 260–263, 304
Methanosaeta 244–246, 248–249,
  252, 263
Methanosarcina 244–246, 248–249,
  263
Methanothermobacter 244–246, 249
microalgae 293–298, 300–310
Microbacteriaceae 212

Microbacterium 140–143, 146, 149,
  157, 207–208, 211
microbial diversity 134, 152, 156, 160,
  165, 239, 244, 261–262
Micrococcaceae 212
microorganisms vii, xix, 3–5, 7, 9, 11,
  13, 15–17, 19, 21, 43, 70, 105–106,
  109, 113, 116, 126, 135, 137–138,
  151, 168–169, 175, 179, 193, 205–
  206, 208, 210, 214, 216, 244, 260,
  262, 270, 275, 293, 305, 308–309
microstructure 4, 15, 17
mineralization 69, 157, 211, 234, 248
model parameters 24, 86
models i, iii, vii, xix, xxii, 1, 4, 11,
  15, 17, 19, 23–26, 28, 30, 34, 36–39,
  42–47, 69, 72, 78–80, 84–87, 93,
  96–98, 111, 131–132, 135, 195, 215,
  220–224, 231–233, 235, 256, 258,
  260, 265, 272
  computational models 15, 19
Moraxellaceae 207, 212

N

Neisseria 270, 276
nitrate 28, 34, 43, 77–78, 80, 83, 106–
  109, 111–113, 115–119, 125–126,
  128, 130–134, 160, 177, 185, 194,
  294, 298, 308–309
  nitrate reduction 34, 112
nitrification xix, 31, 42–45, 47, 132,
  160, 165, 167–171, 176, 182, 194,
  196, 293–295, 301, 303
nitrite 28, 31, 36, 42–46, 131, 133,
  160, 168, 170, 176–178, 183, 185,
  188, 194–196
Nitrobacter 160
Nitrosococcus 160, 167, 170
Nitrosomonas 160, 166–168
Nitrospira 157, 160, 165, 167, 170
Nocardiaceae 212

numerical optimal approaching procedure 25
NOAP 25, 27, 29, 32, 34, 36–39,
42
nutrients vii, ix, xix, xxi, 23, 45, 50,
71, 105–109, 111–113, 115–117, 119,
121–123, 125, 127–131, 133, 135, 138,
142–143, 147, 149–151, 153, 159,
177–178, 189, 192, 196–197, 211, 215,
225, 244, 262, 283, 293, 295, 297–301,
303, 305–311

O

orbital energy levels 20
organic matter xix, 21–22, 64, 69,
71–73, 219, 234, 248
oxidation state 10

P

Paracoccus 122, 125, 128, 135, 168
parameter
estimation 25–27, 34, 37–39, 43,
45, 47
parameter identification 24
particulate organic carbon 52
TOCp 52, 55–62, 68
pH viii, xxi, 5, 9–11, 14, 16–19, 21,
68, 72, 77, 81–82, 91–93, 95, 98,
100, 106, 108, 114, 116, 128, 131,
133–135, 152, 154–155, 177–179,
181–183, 185–186, 189, 192–196,
237–239, 241, 243, 245–249, 251,
253–255, 257, 259, 261, 263–265,
284, 303, 305, 307, 310
phosphate accumulating organisms
106, 131, 195–196
PAOs 178–179, 181–182, 188–189,
193, 196
phosphorous removal
A/O 177–194

O/EI 177–194
photosynthesis 293, 304, 308 310
Planctomycetes 157, 164–165
polyhydoxybutyrate 106
PHB vii, 105–109, 111, 113, 115,
117, 119–123, 125–135, 194
polyhydroxyalkanoate 105, 131
PHA 105–106, 117, 131, 133–134,
180, 184, 196
polyhydroxyvalerate 106
PHV 106, 194
polysaccharides 13, 70, 238, 308
potentiometric titration 5, 18
proteins 64, 70, 208, 238, 244, 250,
259
Proteobacteria 122–123, 141–142,
146, 157, 161, 163, 165, 169, 171,
242, 244, 247
Pseudomonadaceae 125, 276, 281
Pseudomonadales 125
Pseudomonas 20, 128, 135, 140–143,
146–147, 149, 157, 168, 171, 215–
216, 264, 270, 276–278, 291–292
pyrosequencing viii, 110, 134–135,
159–162, 164–172, 239–240,
261–262

Q

quorum sensing ix, 138, 269–271, 275,
277, 283, 290–292
QS 270–272, 275, 278, 285, 319

R

radial structure function 11
RSF 11–12
rarefaction analysis 121
rDNA 110, 119–120, 122, 128,
205–207, 261
relative hydrophobicity 78, 80
RH 78, 80, 88–89

renewable energy 237–238, 253
Rhodobacterales 122, 125
Rhodobacterececae 122
Rhodocyclaceae 125–126
Rhodocyclales 125
ribosomal database project 110, 132
    RDP 110–111, 122, 125, 130, 162
rRNA viii, 132, 134, 140–141, 146,
    152, 154–155, 157, 159–165, 167,
    169, 171, 194, 263, 271, 275–277,
    281, 286–287, 290

S

saturation 4, 79, 97, 208, 293, 308
sedimentation 49–51, 56–58, 61–62,
    68, 71, 153, 253
sensitivity analysis 24–26, 30, 33,
    36–38, 40–41, 46–47
sequential batch reactor 25
    SBR 25, 28–31, 33–35, 37, 39, 43,
    108–110, 122, 130, 133
sewage
    freshwater sewage vii, 75, 77–79,
    81, 83, 85, 87–95, 97, 99
    urban sewage 165
Shewanellacae 212
sludge
    autoclaved sludge 238
    primary sludge 50–52, 56–57, 68,
    245
sludge retention time 43, 131, 192,
    295–296, 299
    SRT 43, 50, 68–69, 106, 131, 192,
    295–296, 299
sludge volume index 52
    SVI 52, 55–56, 68
SMX 138–154
    SMX biodegradation 138–145,
    147–150, 153
    SMX elimination 138, 147, 150

solids retention time 50–51, 71, 171
    SRT 43, 50, 68–69, 106, 131, 192,
    295–296, 299
Spirochaetes 164–165
starvation 51, 59–60, 70, 307, 310
sulfamethoxazole viii, 100, 137–139,
    141, 145, 147, 149, 151, 153,
    155–158
sulfide 225, 228–229, 235
surface complex constant 8
surface complexation 3, 8, 21
synthetic 16, 20, 68, 81, 90, 107–108,
    132–133, 191–193, 253, 258, 284,
    290

T

tetracycline vii, 75–83, 85–87,
    89–101, 155–156, 283–285, 288
Thaurea 126
thermodynamic stability 8, 21
titration curve 5–6
TM7 123, 128, 165
TM7_genera_incertae_sedis 125
total suspended solids 50, 52
    TSS 50, 52–56, 60, 63, 65
toxic xix, 159, 216
Truepera 165

U

UV absorbance 55, 64, 70, 72

V

Variovorax 140–142, 146, 149, 157
Verrumicrobia 164–165
viruses 241, 245
volatile suspended solids 16, 52, 302
    VSS 16, 52, 54, 56, 67, 109, 184,
    240, 243, 255, 299, 302

volatilization 76–78, 83, 97

W

waste activated sludge viii, 49, 71,
98, 219, 223, 230, 233–235, 237,
239, 241, 243, 245, 247, 249, 251,
253–255, 257, 259–261, 263, 265
WAS x, xxii, xxiv, 4–5, 7–11, 13,
15–20, 31, 43, 49–62, 64, 67–70,
76–83, 87–91, 93, 95–98, 106–
113, 115–117, 119–122, 125, 128,
130, 138–140, 142–144, 146–147,
149–155, 161–162, 165–169, 176,
179–183, 185, 188–189, 191–194,
202–208, 210–212, 214, 219–220,
222–223, 225, 227–232, 237–246,
248–251, 253–255, 257–259, 271,
275–280, 284–289, 295, 297,
299–301, 303–308, 313–319
wastewater vii–ix, xix, 3, 5, 15–16,
20, 22–24, 30, 43–47, 49–53, 55–63,
65, 67–73, 76, 78, 81, 98–100, 105,
107–108, 115, 125, 131–135, 138,
151, 155–156, 159–161, 163, 165,
167–172, 175–177, 179, 181, 183,
185, 187, 189–193, 195–197, 204,
219, 225, 228, 230, 233–235, 237,

251, 253, 258, 260, 262, 264, 269–
271, 275, 284, 287, 290, 292 301,
303–311
    domestic wastewater 46, 176, 190,
    195, 233
    industrial wastewater 23, 135, 151
    saline wastewater 81, 167, 171, 197
wastewater treatment systems xix, 24,
43, 45, 171, 225, 294, 305
    WWTP 37, 43, 46, 77, 89, 105,
    123, 126, 149, 152, 160–161,
    165–167, 192–193, 204, 220, 225,
    227–228, 239, 295, 297, 299,
    302–303, 305–306

X

X-ray Absorption Fine Structure 4
    XAFS 4, 9–10, 13, 15, 17–19, 21,
    313
XANES spectra 9–11
xenobiotics viii, 149, 199, 202–203,
    211, 250

Z

zeta potential 78–79, 88–89
ZVI 220, 222–223, 225–231